MINERALS

A GUIDE TO FAMILIAR MINERALS, GEMS, ORES AND ROCKS
a Golden Guide® from St. Martin's Press

by HERBERT S. ZIM and PAUL R. SHAFFER

Revised by
JONATHAN P. LATIMER
and
KAREN STRAY NOLTING
with
JEFFREY POST and PAUL W. POHWAT
NATIONAL MUSEUM OF NATURAL HISTORY

Illustrated by
RAYMOND PERLMAN

St. Martin's Press 〽 New York

FOREWORD

GOLD CRYSTAL

For more than a generation this book has introduced thousands of children and adults to the diversity of the natural world. This guide illustrates hundreds of the most familiar rocks, gems, and minerals, and the text gives concise information to help identify each one. This revision was edited with the assistance of experts from the National Museum of Natural History at the Smithsonian Institution. It reflects the latest information on minerals.

The original edition of this book relied on the cooperation and helpful comments of many experts including Drs. Chapman, Droste, Grim, Hagner, Henderson, Merrill, Wanless, and White of the Geology Department at the University of Illinois, and Drs. Collinson, Swann, and Willman of the Illinois State Geological Survey. The original data for the maps and other factual information in the book came from the U. S. Bureau of Mines and the U. S. Geological Survey.

Special thanks go to Herbert S. Zim, who conceived these guides and coauthored this guide as well as the original edition, entitled *Rocks and Minerals,* and to Paul R. Shaffer, who coauthored the original edition of this guide.

J. P. L.   K. S. N.

ROCKS, GEMS AND MINERALS. Copyright © 2001 by St. Martin's Press. All rights reserved. Printed in China. No part of this book may be used or reproduced in any manner whatsoever without written permission except in the case of brief quotations embodied in critical articles or reviews. For information, address St. Martin's Press, 175 Fifth Avenue, New York, N.Y. 10010. www.stmartins.com

First published under the title *Rocks and Minerals*.

A Golden Guide® is a registered trademark of Golden Books Publishing Company, Inc., used under license.

ISBN 1-58238-132-1

CONTENTS

HOW TO USE THIS BOOK

This is a guide to aid you in identifying rocks and minerals. But it is more than that. This book will also help you understand the importance of rocks and minerals in our daily lives. Hence you will find information on the uses of minerals and mineral products as well as aids in identification.

Skill in identifying rocks and minerals comes with experience. Take this book on hikes, trips, and vacations. Visit collecting places, examine specimens, and try simple field tests. Remember that only the most common kinds of rocks and minerals are shown in this book. Even these may vary considerably in their physical properties.

Thumb through this book before you go out on trips. Read the introductory sections. Next, become familiar with the pictures of minerals, rocks, and rock structures. This may enable you to identify some common rocks and minerals on sight. Maps on some pages show where important deposits are located. Books listed on p. 156 will help further, but local inquiry is often needed to pin-point locations.

As you make observations and collect specimens, check your book or make marginal notes for later reference use. In the long run, your records may be as important as the specimens you collect.

4

THE EARTH AND ITS ROCKS

SEEING ROCKS AND MINERALS is something hard to avoid. You have to stalk wildlife and await spring flowers, but every roadcut, bank, cliff, excavation, or quarry exposes rocks and minerals for you to see. Learn to watch for *outcrops*, places where bedrock is naturally exposed, as in ridges or cliffs. Wherever erosion is taking place, rock is sooner or later exposed. Outcrops of rocks are common in many parts of New England and the Southwest. Where soil is thick, as in the middle states, one may have to search for outcrops along river valleys and on steeper hillsides.

Man-made exposures of rocks and minerals are often the best source of specimens. Look in road and railroad cuts, in quarries, rock pits, dump piles around mines, and similar places. Look with care. Ask permission before entering a quarry. Be alert for new roads, bridges, or foundations where excavations expose fresh rocks.

SURFACE FORM	ROCK	MINERAL
lava-capped butte	made of rock— limestone	contains mineral— calcite

Look for Surface Forms These are evident in the physical forms seen on the surface—hills, valleys, cliffs, and basins. Some forms and structures are associated with a particular type of rock and can be recognized at a distance. Knowing forms and structures does more than help you find rocks and minerals. It aids you in interpreting the landscape, so that you can read dramatic chapters of the history of the earth. Rock structures are dealt with under rocks (pp. 121–145). Physical geology treats them in greater detail.

Look for Rocks These are the materials of which the crust of the earth is made. They form the mountains and underlie the valleys. You see them when they have been pushed or folded upward or when they jut through soil to form an outcrop. Minerals are the components of rocks. Rocks include the solid bedrock and also the unconsolidated debris above it, called the mantle.

Look for Minerals These are the building blocks of which the earth itself is made. They are inorganic chemical elements or compounds (some quite complex) found naturally in the earth. Because each is a chemical compound, a mineral has fairly definite and stable properties.

What Is Behind it All is as mysterious as the origin of the universe, the solar system, and the earth. Geology is the science of the earth and its history. Mineralogy (the study of minerals) and petrology (the study of rocks) are two divisions of the science of geology.

The Earth is a ball of rock 7927 miles in diameter, weighing about 6.6 sextillion tons. It is composed of a dense iron-nickel core, a molten outer core, and several massive rock layers—denser near the core and lighter toward the surface. The continents themselves are islands of granitic rock (p. 111) floating on the denser, darker rocks which underlie the ocean basins and go down perhaps 40 miles. Most of our economic minerals are in the lighter granitic rocks and rock derived from them. In the process of mountain building some of the darker rocks have come to the surface and can be seen.

The Earth and Man are inseparable, even during an age of space travel. It is impossible to name a major industry which is not directly or indirectly dependent upon rocks and minerals. Mineral resources are extremely important, and the increased use of them is a direct measure of progress. It is impossible to think of a time when dependence on minerals will cease. New uses for minerals open technological vistas. Uranium minerals, once oddities, are now a source of energy.

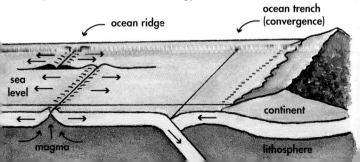

ocean trench (convergence)

ocean ridge

sea level

magma

continent

lithosphere

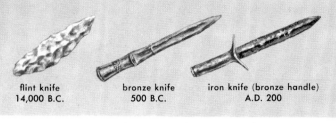

| flint knife | bronze knife | iron knife (bronze handle) |
| 14,000 B.C. | 500 B.C. | A.D. 200 |

Early man made these weapons by using mineral resources.

Mineral Resources were important thousands of years ago when men trekked hundreds of miles to flint quarries. As men learned to make bronze and steel, they became essential. The fossil fuels—coal and petroleum—are the ones on which we still depend, and the nuclear industry needs uranium, thorium, and similar metals.

Conservation of mineral resources poses real problems, for most of these can never be replaced. Conservation of mineral specimens is a special case. In some single deposits of rare minerals, the first collectors have taken the entire supply. Careless collectors often spoil fine material. When you go collecting, remember to be considerate of others.

Limestone quarry near Bloomington, Ind.

Riker mount

Mineral cabinet

ACTIVITIES FOR AMATEURS

Collecting is the first and most obvious activity for anyone interested in rocks and minerals. And well it might be, for the number of mineral species comes to about 4000, and there are hundreds of kinds of rocks. A collection enables you to study, compare, and analyze minerals; hence you learn more about them. It's fun to find, buy, and swap specimens. Collecting takes you out-of-doors; it also paves the way for more serious studies in science or engineering.

Where to Collect In addition to the general places mentioned on p. 5, run down specific mineral localities near you. Become familiar with the books and magazines listed on p. 156 and with publications of your state geological survey. Mineralogical magazines often list mineral localities, and in many states guides to mineral deposits are available. Ask mineralogists and local collectors. They will be glad to help you out.

How to Collect involves the place you go to. A collecting trip should have specific objectives. Study the area in advance to learn the lay of the land, accessibility, rock structure, and possible minerals. Allow ample time. Suit your plans to the location. In deserts, collect early or late; midday sun on bare rock can be overpowering. Work systematically; don't attempt too much.

geologist's hammer

old newspapers

cold chisel

field bag

magnifying glass

compass

notebook and pencil

heavy gloves

Equipment can be simple: a pile of old newspaper for wrapping specimens, a notebook and pencil, are all you need at a mine dump, where material is broken. Otherwise, a geologist's pick or plasterer's hammer is essential—and get a good one. A cold chisel, a magnifying glass, compass, heavy gloves, a pocket knife, and a shoulder bag or knapsack are useful. Don't carry too much. You will need water and lunch—and a knapsack full of rocks gets heavy.

The Specimens You Collect should be selected with care. From a dozen that seem likely, keep only a few. Hand-sized specimens are preferred. Some collectors look for small, perfect, thumbnail-sized specimens and study them with a low-power microscope. Be sure your specimen is fresh. Trim it to shape. This takes skill—you soon learn that light blows placed correctly do the trick. Wrap your specimens as shown below. Include a label with field identification, location, and date.

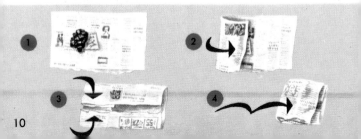

SPECIMENS AT HOME take up room and gather dust. Collecting is fun, but give some thought to what you will do with your collection. Some rocks and minerals are so attractive that you may want them for their sheer beauty. Set these where you can see and enjoy them. But if you intend to have a study collection, more is required. As a first step each specimen must be accurately identified, classified, and labeled.

Identification is best made by the physical characteristics of minerals and rocks (pp. 13-24). Check the hardness, streak, specific gravity and, possibly, the crystal form. Often the geologist at the state university, museum, or geological survey will identify difficult specimens. Send him a small sample with exact details of where it was found.

Classification of rocks and minerals depends on your purposes. In this book for beginners, a very simple classification is used. More advanced books follow a standard chemical classification for minerals, and a more detailed classification of rocks. Classification goes one step beyond identification. It shows you the relationship between rocks or minerals. That points the way to discovering their origin, history, and modes of occurrence.

Labeling usually begins with a spot of quick-drying enamel put on the specimen in an inconspicuous place. After it dries, a number is added in India ink. This number refers to a label and a catalog, both of which should include number, name, location, collector, date, and rock or mineral associations.

No. 41   Date 9/13/58
Name Hematite
Locality Rome, N.Y.
Collector H.S. Zim

AGATE   rough   cut sections

shaped sections   polished   mounted

OTHER ACTIVITIES related to rocks and minerals include the art and science of gem collecting and gem cutting. An amateur can collect or purchase an abundance of semiprecious stones which he can cut and polish (p. 90) to bring out their brilliance. From this it is only one step to making your own jewelry. Photographing rocks and minerals is a challenge to the cameraman. A close-up lens reveals striking details of form and color. Finally, more experienced amateurs may want to specialize in the minerals of their home locality or in certain groups, such as the quartz minerals. Others may find that experiments with fluorescent (p. 22) or radioactive minerals are important enough to demand their undivided attention.

Mineralogy Clubs are so numerous that it is impossible to list them here. Some clubs specialize in gems, gem cutting, and polishing, but most amateur groups are concerned with all aspects of rocks and minerals. Some are for all, including beginners; a few are for advanced students and professionals. All have meetings, exhibits, field trips, and sometimes their own laboratories. Join a club if there is one nearby. There is no better way to get started.

JADEITE is sodium aluminum silicate $(Na(Al,Fe)Si_2O_6)$. Color: white, yellow, brown, or green. Often a gem (p. 88); seldom as crystals.

SPODUMENE is lithium aluminum silicate $(LiAlSi_2O_6)$; opaque—white, lilac, or yellow. Rarely a transparent crystalline gem stone.

IDENTIFYING MINERALS

MINERALS ARE CHEMICALS They are chemical elements or compounds found naturally in the crust of the earth. They are inorganic, in contrast to organic chemicals (made mainly of carbon, hydrogen, and oxygen) typical of living things. Some minerals have a fixed chemical composition. Others are a series of related compositions where one metallic element may wholly or partly replace another. The two minerals above are very similar chemically and in some of their physical properties, but are usually quite different in color and other physical properties. Only rarely will a single physical or chemical property identify a mineral. Usually more characteristics must be used. These physical and chemical properties are described on pp. 14 to 24. Some are inherent and reliable, others are variable and must be used with care. You can easily learn to use the simpler physical and chemical tests. Identification of many rare minerals often requires expensive laboratory equipment and detailed chemical and optical tests which only an expert can make.

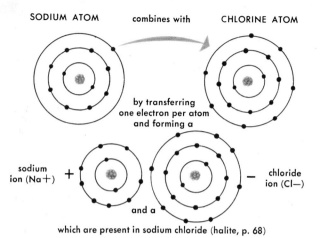

SODIUM ATOM   combines with   CHLORINE ATOM

by transferring
one electron per atom
and forming a

sodium
ion (Na+)   +   chloride
ion (Cl−)

and a

which are present in sodium chloride (halite, p. 68)

ELEMENTS are the building blocks of all materials, including minerals and rocks. About 88 naturally occurring elements are known. A dozen or so were known in ancient times; the latest were found in atom-splitting experiments. All are made up of protons, neutrons, and electrons. These, combined, form atoms of matter. The atoms in turn join to form molecules—the smallest particles usually produced in chemical reactions. When temperatures are high, molecules may break down into atoms or atom groups. With slow cooling these may join together, in regular order, to form crystals. Most minerals are crystalline, being formed from cooling melts, solutions, or vapors in the earth. The arrangement of an atom's electrons determines with what other elements it will combine, and in what proportions.

INTERFERENCE COLORS are shown by this thin section of a rock rich in pyroxene, here viewed through a polarizing microscope.

BIAXIAL INTERFERENCE FIGURE forms when a thin sheet of muscovite mica is examined through a polarizing microscope.

OPTICAL PROPERTIES of minerals are used mainly by experts, but amateurs should know about them because they are fundamental in precise mineral identification. Optical identification is highly accurate and can be used with particles of microscopic size. X-rays sent through thin fragments or powders produce a pattern dependent on the atomic structure and so are an aid to identification. Pieces of minerals or rocks are mounted on slides, then ground till paper thin. These thin sections are examined through ordinary and polarized light. The bending of light as it passes through the minerals gives patterns that aid in identification. Fragments of minerals can be immersed in transparent liquids of different density to measure their index of refraction. This is distinct for each mineral and is related to its crystal system (pp. 16–17). Thus an expert can tell if a diamond or emerald is real or false without doing any damage to the stone. It is worth paying more attention to these optical properties as you become more experienced.

A check on this diamond's index of refraction shows that bottom is cemented glass.

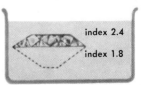

index 2.4

index 1.8

CRYSTAL FORM is critical in mineral identification as it reflects the atomic structure of the mineral. It is also the most difficult characteristic to use and one that requires the most careful study. Yet crystals are so magnificent in their beauty and symmetry it is sometimes hard to believe they are natural. The study of how they are formed reveals mathematical relationships as amazing and as beautiful as the crystals themselves. Perfect crystals are rare, and some

GALENA   HALITE

Cubic (Isometric) System includes crystals in which the three axes (common to five of six systems) are of equal length and are at right angles to one another, as in a cube— examples: galena, garnet, pyrite, and halite.

ZIRCON   RUTILE

Tetragonal System has two axes of equal length and one unequal. All three axes are at right angles to one another, as in zircon, rutile, and cassiterite.

QUARTZ   CALCITE

Hexagonal System has three equal axes at 120° angles arranged in one plane and one more axis of a different length at right angles to these, as in quartz, beryl, calcite, tourmaline, and cinnabar.

are of great value. However, a fragment or an imperfect crystal will yield basic data to the experienced mineralogist. Sometimes crystals develop in clusters, or as twins. They reveal distortions, inclusions, and other interruptions in their development. A very simple outline of the six systems of crystals is given on these two pages as a bare introduction to the science of crystallography:

Orthorhombic System has crystals with three axes all at right angles, but all of different length. Examples: sulfur, barite, celestite, staurolite, and olivine.

Monoclinic System has three unequal axes, two of which are not at right angles. The third makes a right angle to the plane of the other two, as in orthoclase, gypsum, micas, augite, epidote, and hornblende.

Triclinic System has three unequal axes but none forms a right angle with any other. Examples: plagioclase feldspars, rhodonite, and chalcanthite.

SULFUR   STAUROLITE

EPIDOTE   AUGITE

AMAZONSTONE   RHODONITE

A scratches B

B does not scratch A

Try to scratch
A with B

Try to scratch
B with A

HARDNESS is used in a rough manner in mineral identi-fication. There are much more precise ways of measuring hardness in industrial laboratories. Though arbitrary, Mohs' scale of ten minerals is useful:

1. Talc	6. Orthoclase
2. Gypsum	7. Quartz
3. Calcite	8. Topaz
4. Fluorite	9. Corundum
5. Apatite	10. Diamond

Gypsum is harder than talc but not twice as hard; fluorite is harder than calcite but not as hard as apatite. If an unknown material will scratch all the minerals in the scale up to 4 and is scratched by apatite, its hardness is between 4 and 5. Check carefully to be sure there is a distinct scratch, but never test hardness on the face of a valuable crystal. For field use here are some other convenient stan-dards of hardness:

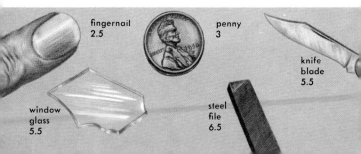

fingernail
2.5

penny
3

knife
blade
5.5

window
glass
5.5

steel
file
6.5

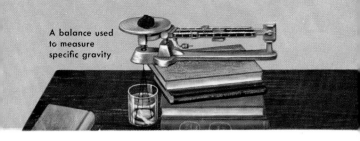

A balance used to measure specific gravity

SPECIFIC GRAVITY is the relative weight of a mineral compared to the weight of an equal volume of water. Since the weight of an equal volume of water is identical with the mineral's loss in weight when weighed in water, specific gravity (Sp. Gr.) is quickly determined. A corundum crystal weighing 2.0 oz. dry weighs 1.5 oz. when suspended in water. The loss (0.5 oz.) divided into the dry weight gives a specific gravity of 4.0. This may seem odd, because corundum contains only aluminum (Sp. Gr. 2.5) and oxygen, a gas. Learn to estimate specific gravity and make your own measurements as an aid to identification. Below are some average figures.

Borax	1 7	Talc	2.8	Corundum	4.0
Sulfur	2.0	Muscovite	2.8	Rutile	4.2
Halite	2.1	Tremolite	3.0	Barite	4.5
Stilbite	2.2	Apatite	3.2	Zircon	4.7
Gypsum	2.3	Crocidolite	3.3	Zincite	5.5
Serpentine	2.5	Topaz	3.5	Cassiterite	7.0
Orthoclase	2.6	Rhodochrosite	3.6	Cinnabar	8.0
Quartz	2.7	Staurolite	3.7	Uraninite	9.5
Calcite	2.7	Siderite	3.9	Gold	19.3

(Sp. Gr. of some minerals may vary as much as 25 per cent from specimen to specimen.)

aluminum
Sp. Gr. 2.5

+

oxygen
a gas

=

corundum
Sp. Gr. 4.0

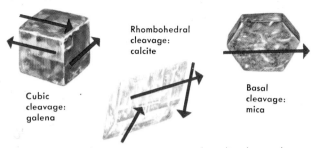

Cubic
cleavage:
galena

Rhombohedral
cleavage:
calcite

Basal
cleavage:
mica

CLEAVAGE is the way some minerals split along planes of weakness in the atomic structure of the mineral and parallel to possible crystal faces. The perfection of cleavage is described in five steps from poor (as in bornite) to fair, good, perfect, and eminent (as in micas). The types of cleavage are usually described by the number and direction of cleavage planes. Three examples of cleavage are shown above. Use cleavage as an aid in identification—though you may at first find it difficult to tell the face of a crystal from a fresh, perfect cleavage surface.

Fracture is the breakage of a mineral specimen in some way other than along cleavage planes. Not all minerals show good cleavage; most show fracture. Fresh fractures show the mineral's true color. Five to seven types of fracture are recognized; three are shown below.

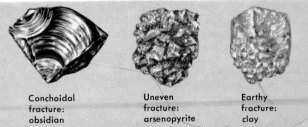

Conchoidal
fracture:
obsidian

Uneven
fracture:
arsenopyrite

Earthy
fracture:
clay

COLOR is the first of three characteristics that have to do with the way a mineral looks. In most metallic ores it is a safe clue in identification. But in quartz, corundum, calcite, fluorite, agate, garnet, tourmaline, and others it is often due to impurities and may vary greatly. So use color with caution and use only the color of a freshly broken surface. Note the surface tarnish on some metallic ores; it differs from the true color, but can be used for identification also.

Streak is the color of the powdered mineral best seen when the mineral is rubbed against a streak plate of unglazed porcelain (the back of a tile is excellent). In metallic ores the streak may differ from the color and so is worth noting.

Luster depends on the absorption, reflection, or refraction of light by the surface of a mineral. It is often an aid in identification. About a dozen terms are used, most of which are self-explanatory: *adamantine* (brilliant), like diamond; *vitreous* (glassy), like quartz; and *metallic* (like metal), like galena. The prefix *sub-* is used when the characteristic is less clear. Other lusters to note: dull, earthy, silky, greasy, pearly, resinous.

TOURMALINE CRYSTAL

CORUNDUM

USING A STREAK PLATE

LUSTER

metallic

resinous

silky

glassy

FLUORESCENT LIGHT

NATURAL LIGHT

WILLEMITE
AND CALCITE

SCHEELITE

COMMON
OPAL

ULTRAVIOLET LIGHT is invisible. Its waves are too short to be detected by the eye. However, some minerals, when exposed to this light, are "excited"—they absorb the ultraviolet light and emit longer light rays which we see as colors. Minerals which do this are *fluorescent.* If they continue to emit light after the ultraviolet rays have been cut off, they are *phosphorescent* (like the luminous dial of a watch). A quartz lamp is a fine source of ultraviolet light of short (about 1/10,000 in.) wavelength. An argon light gives longer ultraviolet rays. Not all minerals fluoresce when exposed to ultraviolet light. Some uranium minerals do; so does scheelite, an ore of tungsten. Other minerals may fluoresce because of impurities. Some fluoresce when they come from one locality and not another. This makes the search for fluorescent minerals exciting. A portable ultraviolet lamp can be used on field trips. Because of their beauty, fluorescent minerals receive a good deal of attention, but the serious study of fluorescence is a difficult one.

MAGNETISM occurs in a few minerals. Lodestone (a form of magnetite) is a natural magnet. An alnico magnet will attract bits of magnetite and pyrrhotite. A few manganese, nickel, and iron-titanium ores become magnetic when heated by a blowpipe.

Pyrrhotite
is magnetic

ELECTRICAL PROPERTIES of minerals are varied. Thin slabs of quartz crystal are used in watches and clocks. Crystals of sulfur, topaz, and other minerals develop an electric charge when rubbed. Tourmaline crystals, when heated, develop opposite charges at opposite ends of the crystal.

Heated tourmaline
develops electric
charges

HEAT may raise the temperature of a mineral till it will fuse in a blowpipe flame. Use only small, thin splinters. The seven-point scale of fusibility, with examples, is:

1. Stibnite: fuses in alcohol lamp or candle flame — (980°F)
2. Chalcopyrite: fuses easily in blowpipe flame — (1475°F)
3. Almandite: fuses less easily in blowpipe flame — (1920°F)
4. Actinolite: thin edges fuse easily with blowpipe — (2190°F)
5. Orthoclase: thin edges fuse with difficulty — (2374°F)
6. Enstatite: only thinnest edges fuse with blowpipe — (2550°F)
7. Quartz: no fusing at all in blowpipe flame — (Over 2550°F)

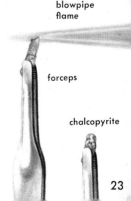

blowpipe
flame

forceps

chalcopyrite

23

GEIGER COUNTERS are not typically a tool for the amateur mineralogist but have become well known because of their use in locating or checking deposits of uranium and thorium. The Geiger tube is the heart of the counter. A wire down the center of the tube has a negative charge. However, no current flows, because of the gas that fills the tube. When the tube is exposed to radiation, some of the molecules of the gas are ionized. They develop an electric charge because of electrons knocked from them. The ionized gas conducts electricity and there is a momentary flash of current through the tube. This is recorded on a counter or dial, or heard as a click in headphones. Cosmic rays from outer space also discharge the Geiger tube, and these must be considered when searching for ore. In order to get an effect, the ordinary Geiger counter must be held close to the ore. Hence it works best on outcrops and other exposed rock. The richer the ore, the more numerous the discharges and clicks. The scintillometer is another sensitive apparatus used to detect radioactive materials.

MINERALS as natural inorganic chemicals can be identified by their chemical properties as well as by hardness, streak, or luster. Once a mineral is dissolved (often a difficult task), other chemical solutions can be added to identify the elements in it. Chemists make constant use of these "wet" tests. Although prospectors working in isolated places can frequently identify a mineral or rock by its physical properties, they also collect and bring material in for laboratory study.

Laboratory tests often involve a blowpipe, a short metal tube for blowing air into a flame. In blowpipe analysis, a bit of the mineral to be tested is heated on a charcoal block. The colored coatings which form identify the elements present. The mineral may be powdered, also, and a touch of the powder absorbed in a drop of melted borax to give a bead test (p. 28). When mineral powder is brought into a flame, the flame color may be studied. Powdered minerals are also heated in closed or open tubes (p. 30).

Mineral analysis calls for the basic materials listed below—as well as for a working knowledge of chemistry. Books to help you are listed on p. 156.

Blowpipe
Charcoal blocks
Alcohol or gas burner
Test tubes
Open tubes
Chemical forceps
Platinum wire

Mortar and pestle
Blue and green glass
Hammer and small anvil
Litmus paper
Borax powder and other
   chemicals
Magnet

REDUCING FLAME

OXIDIZING FLAME

Use blowpipe to send a stream of air into flame, molding it into a narrow cone about 2 in. long. For reducing flame, hold blowpipe behind flame and heat specimen at tip of interior bluish cone. For oxidizing flame, hold blowpipe in flame and heat specimen at tip.

BLOWPIPE TESTS make use of an alcohol, gas, or candle flame. Moving the blowpipe back and forth as shown above produces an oxidizing flame (extra oxygen comes from the air blown into the flame) or a reducing flame (hot gases take oxygen from the specimen). The powdered specimen set in a hollow at one end of a charcoal block is heated in the flame until changes occur. Sometimes a fine, colored coating (sublimate) forms; sometimes a bead of metal is left behind or characteristic fumes are released. When a flux (iodide, bromide, or chromate) is added to the powdered mineral, differently colored coatings form. These may be used to confirm the tests. In some tests the mineral is heated on a block of plaster instead of a charcoal block. Finally, the blowpipe can be used to heat a mineral specimen directly, as in determining fusibility. A sliver of the mineral is held by forceps directly in the flame, as shown on p. 23.

Blowpipe test for zinc: heating ore on charcoal with oxidizing flame.

SAMPLE BLOWPIPE TESTS for the metallic element in the mineral specimen:

Antimony forms a dense white coating, bluish at the fringe; it is volatile, but not as much as arsenic, which forms a similar coating and has a garlic odor.

ANTIMONY

Bismuth minerals give an orange-yellow coating, which becomes greenish-yellow on cooling. A gray brittle button of bismuth also forms. When iodide flux is mixed with the powdered mineral, the coating is yellow with a reddish border.

BISMUTH

Copper When oxides are heated in the reducing flame with a flux practically no coating results, but a reddish ball of metallic copper remains. Blowpipe flame is colored blue-green.

COPPER

Lead minerals heated in a reducing flame leave behind a gray ball of metallic lead. The coating is yellowish (darker when hot) with a white or bluish border.

LEAD

Zinc minerals give a small coating close to the specimen—bright yellow when hot (p. 26), white when cold. Add a drop of cobalt nitrate solution to the coating, reheat, and the coating will turn green. Use plaster block with iodide flux.

ZINC

27

Bead in oxidizing flame.

BEAD TESTS help identify metals when minerals are dissolved in a flux and heated. The flux is borax, heated in a loop of platinum wire till a clear glassy bead is formed. The hot borax bead is touched to a trace of the powdered mineral and is reheated in reducing and oxidizing flames. The color of the bead is noted when hot and when cold. All traces of a previous bead must be "washed" off the wire before a new test is made. Below are the colors in some common bead tests.

	OXIDIZING FLAME		REDUCING FLAME	
Metal	**Hot**	**Cold**	**Hot**	**Cold**
Antimony	yellow	colorless	yellow	colorless
Chromium	yellow	green	green	green
Cobalt	blue	blue	blue	blue
Copper	green	blue	colorless	brown
Iron	yellow	green	green	green
Manganese	violet-brown	violet	colorless	colorless
Molybdenum	yellow	colorless	brown	brown
Nickel	violet	brown	colorless-gray	colorless-gray
Titanium	colorless	colorless-white	yellow-gray	yellow
Tungsten	yellow	colorless	yellow	brown
Uranium	yellow	yellow-brown	green	green
Vanadium	yellow	green	brown	green

After M. Zim, Blowpipe Analysis and Tests for Common Minerals, *1935.*

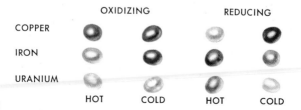

	OXIDIZING		REDUCING	
COPPER				
IRON				
URANIUM				
	HOT	COLD	HOT	COLD

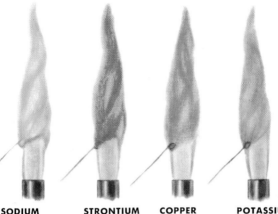

SODIUM
Strong yellow, invisible through cobalt glass.

STRONTIUM
Strong, intense crimson-red flame.

COPPER
Blue, with some green, depending on ore.

POTASSIUM
Violet, visible as red through cobalt glass.

FLAME TESTS depend on the fact that small amounts of mineral introduced into a flame will color the flame, depending on the metals it contains. Flame tests are fairly crude, but when such flames are viewed through a spectroscope, a highly accurate kind of chemical analysis is possible. The spectroscope and the X-ray are important tools in advanced work with minerals and are essential in working with small, rare specimens.

To make simple flame tests, use a clean platinum or nichrome wire. The wire, dipped in strong hydrochloric acid, is held in the flame until no change in flame color is seen. The wire loop is touched to a bit of the powdered mineral, also moistened with acid, and then flamed. The flame may be viewed directly, or through a cobalt blue glass, which masks the yellow sodium color. The Merwin screen, made of two overlapping layers of colored plastic, is also used to view flame colors.

SULFUR melts in closed tube test. On continued heating it forms a yellow to reddish-brown sublimate in the cool part of the tube.

LEAD ORE (galena) turns a light color and forms a white sublimate when heated in an open tube.

TUBE TESTS involve heating powdered minerals in closed and open tubes to see what sublimates form in the upper, cooler part of the tube, and to notice fumes and odors from the heated mineral. The closed tube is an ordinary pyrex test tube. Use only enough mineral to barely cover the bottom. Hold at a low angle in Bunsen burner flame and heat until mineral is red hot. Watch for fumes and sublimates. The open tube can be a piece of straight or bent glass tubing about six inches long. The powdered mineral is inserted about an inch from the lower end, and the specimen is heated while being held at a low angle. Do not tilt higher or the specimen will spill. Air circulates through the open tube and oxidizes the mineral powder. Sublimates form at the cool end. Use a test tube holder for these tests.

NATIVE COPPER
in matrix

crystal

NATIVE COPPER—Michigan

METALLIC MINERALS

The metals are the core of our civilization. Life as we know it would be impossible without them. Here are the minerals from which our most important metals are obtained; they form an interesting group for collectors.

COPPER nuggets were found by ancient man. Later, copper was smelted from its ores. Today it is essential for practically all things electrical and for many other uses. Chile, Peru, Cyprus, Democratic Republic of Congo, South Africa, Russia, and Australia have large deposits. Our deposits are largely in the Southwest. Copper occurs principally in volcanic rocks and in veins. Native copper (H. 2.5 to 3, Sp. Gr. 8.9) is hard to mine. The sulfides and carbonates described on the next page are easier to handle. Some malachite and azurite is cut for ornaments and gems, as is chrysocolla, a copper silicate. Cuprite (copper oxide), a reddish brown mineral, results from oxidation of other copper minerals. Sulfides are black, purple, and yellow. The oxide and native copper are dull red; the carbonates, blue and green.

31

Butte, Mont.

Chalcocite (Cu_2S) is a dark metallic mineral. H. 2.5; Gr. 5.5; streak, gray to black. An important ore, it is found with the other three minerals on this page. Usually occurs in vein deposits; crystals rare.

crystal

Butte, Mont.

Covellite (CuS), found with other copper sulfide ores, forms thin iridescent blue plates, usually tarnished to purple or black. Not as common or as rich in copper as chalcocite. H. 1.5 to 2; Sp. Gr. 4.6; luster, metallic. Occurs as crystals or incrustations.

Butte, Mont.

Bornite (Cu_5FeS_4) is called peacock ore because of its usually shiny, purple tarnish. An important ore, found in veins or scattered in igneous rock. Crystals rare. Bronze-colored when fresh. H. 3; Sp. Gr. 5; streak, black. May contain small amounts of gold and silver.

Beaver Co., Utah

Chalcopyrite ($CuFeS_2$), the most common copper ore, is a brassy, almost golden mineral. May form crystals but is more often found in massive form, in most copper mines. H. 3.5 to 4; Sp. Gr. 4.2; streak, greenish-black; very brittle.

Cuprite (Cu_2O) forms by weathering of other ores and so is more common near the surface. Cubic crystals fairly common. Also occurs as grains and irregular masses. H. 3.5; Sp. Gr. 6; color, reddish-brown; streak, brownish.

Seven Devils, Wash.

Chrysocolla $(Cu, Al)_2H_2Si_2O_5 (OH)_4 \cdot NH_2O$ is found in veins and masses with quartz in most copper mines in the Southwest. Its chief value is as a gem when even-colored and rich in quartz. Color varies, often bluish green; H. 2 to 4; Sp. Gr. 2.

Bisbee, Ariz.

Malachite and Azurite are commonly found together. Malachite $(Cu_2CO_3(OH)_2)$, more common than azurite, is various shades of green. Azurite $(Cu_3(CO_3)_2(OH)_2)$, which is blue, forms crystals more often. Both occur in smooth or irregular masses in the upper levels of mines. Compact, deep-colored stones are cut as ornaments. H. (for both) about 4; Sp. Gr. 3.7 to 4.

MALACHITE—Bisbee, Ariz.

AZURITE Bisbee, Ariz.

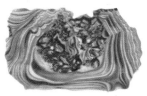

cut and polished

Azurite and malachite mixture

33

GALENA CRYSTALS

LEAD does occur as the native metal, but only rarely. The most important source of lead is the mineral galena, lead sulfide (PbS). It has been known for centuries, and lead, smelted easily from it, has also been used since ancient times. Galena is found in veins, pockets, and replacement deposits in carbonate rocks. It occurs with zinc, copper, and silver, often containing enough of the latter metals to make the ore doubly valuable. About two-thirds of U.S. silver production comes from lead-zinc ores.

Galena is a heavy, brittle, silvery-gray mineral which commonly forms cubic crystals and has perfect cubic cleavage. Its crystals were used in early radio sets. H. 2.5; Sp. Gr. 7.5; streak, lead-gray. Lead deposits are located in southwest Missouri and Tennessee. In Colorado, Idaho, and Utah lead is found with silver ores. Europe, Russia, South America, and Australia also have commercial deposits.

CRYSTALS
AS MINED

cleavage planes

CERUSSITE
Leadville, Colo.

ANGLESITE REPLACING GALENA
Tintic district, Utah

CERUSSITE IN HEMATITE
Embreeville, Tenn.

Over a dozen other lead minerals exist, but of this number only two have much importance as ores. Both are secondary minerals derived from galena by the slow action of air and water. Silver-bearing galena is roasted to form lead oxide and sulfur dioxide gas, then reduced with carbon. Zinc is added. The silver and zinc rise and are skimmed off. The zinc is removed by distillation.

Cerussite ($PbCO_3$) forms large white or gray crystals, sometimes needle-like in bundles. It also occurs as massive deposits or as loose, crystalline crusts. H. 3; Sp. Gr. 6.5; adamantine or silky luster, white streak.

Anglesite ($PbSO_4$) is often found with galena, as a white or gray crust. H. 3, Sp. Gr. 6.4; streak, white

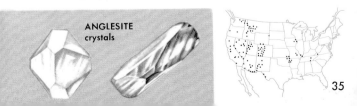

ANGLESITE
crystals

GOLD is neither the rarest nor the most valuable metal, yet it is part of the foundation of trade and commerce, and has many uses because of its metallic properties. Ancients who found native gold prized it, and gold, beautiful and easily worked, is still widely used in jewelry. A soft metal (H. 2.5), it is sometimes alloyed with copper to harden it and make it go further. Pure gold is 24 carats; hence 14 carat gold is 14/24 or about 60 per cent gold. Gold is found in quartz veins, sometimes with pyrite. The gold may occur within the pyrite itself—giving fool's gold a real value. Gold may occur in metamorphic rock and occasionally in sediments where it has been redeposited. Only rarely is visible gold found in gold ore. It usually cannot be seen at all. Some of the commercial ores contain only 0.1 ounce of gold for each ton of rock.

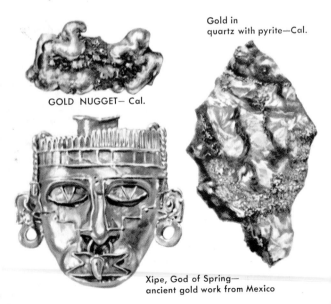

GOLD NUGGET— Cal.

Gold in
quartz with pyrite—Cal.

Xipe, God of Spring—
ancient gold work from Mexico

Mining gold with sluice and pan.

As gold deposits are eroded, the heavy gold (Sp. Gr. 19.3) is concentrated in stream beds where grains, flakes, and even nuggets may be found by washing away the lighter sand in a gold pan or a sluice. From these placer deposits miners have gone on to search for the original veins or "lodes." Here the gold may be found as flecks in the quartz and, rarely, as octahedral crystals. Gold is malleable; color pale to golden yellow; metallic luster. It occurs as a compound with tellurium in such minerals as sylvanite $(Au,Ag)Te_2$ and calaverite, $AuTe_2$. Gold may also be recovered from other metallic ores.

GOLD CRYSTAL 0.1 in.

GOLD ORE—Lead, S. Dak.

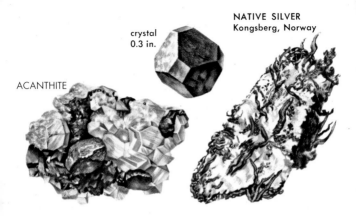

crystal
0.3 in.

ACANTHITE

NATIVE SILVER
Kongsberg, Norway

SILVER sometimes occurs as native silver in large twisting, branching masses. Another important source of silver is the sulfide (Ag_2S)—acanthite. In addition, silver, lead, sulfur, and antimony form a whole series of rare, complex minerals. Silver may be a valuable by-product in smelting lead, zinc, and other metals.

Acanthite is massive, or it may form cubic crystals. Found with lead, copper, and zinc minerals. Color silvery when fresh, black to gray when tarnished. H. 2.5; Sp. Gr. 7.3; luster, metallic.

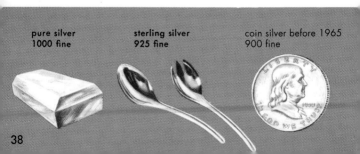

pure silver
1000 fine

sterling silver
925 fine

coin silver before 1965
900 fine

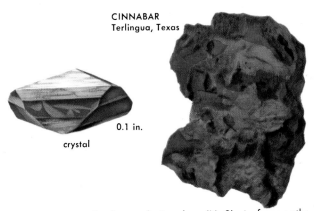

CINNABAR
Terlingua, Texas

0.1 in.

crystal

MERCURY, in the form of cinnabar (HgS), is frequently a bright red, attractive mineral found near hot springs and in low-temperature veins—typically near volcanic rocks. American deposits, concentrated in the Sierra Nevada and coast ranges, are more limited than those of Spain or China. Cinnabar forms hexagonal crystals but is usually massive or occurs as scattered flecks. H. 2.5; Sp. Gr. 8.1; color varies from black to bright red; prismatic cleavage. Mercury is also found as silvery globules of native mercury in deposits of cinnabar. It is used in medicine, in the manufacture of thermometers and explosives, and in several chemical industries.

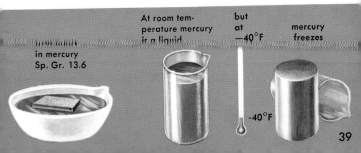

At room temperature mercury is a liquid

but at —40°F

mercury freezes

in mercury
Sp. Gr. 13.6

-40°F

IRON ORES

crystal

Toluca, Mexico

Meteorites are "shooting stars" that reach the earth. Some are stony; some are iron alloyed with nickel and traces of other metals. Meteorites range from sand-grain size to masses weighing tons. Surface often pitted, oxidized, or rusty. Iron meteors are magnetic. H. 4 to 5; Sp. Gr. 7.5.

Hematite (Fe_2O_3) the most important iron ore, contains about 70 per cent iron. Great beds occur in Minnesota, Michigan, Wisconsin, and adjacent Canada. Hematite varies from a red earthy powder to a dark, compact, shiny mineral. H. 1 to 6; Sp. Gr. about 5; streak, cherry red.

Magnetite (Fe_3O_4) is the only black ore that can be picked up easily by a magnet. The hard (H. 6) and heavy (Sp. Gr. 5.2) black crystals or masses are found in basic igneous rocks and metamorphosed sedimentary rocks. A valuable ore, though sometimes difficult to mine. Streak: black.

Goethite (FeOOH) is soft and earthy (called yellow ochre or sometimes limonite), or in compact, smooth, dark, rounded masses, or occasionally as circular crystals. Goethite is about 60 per cent iron. H. 1 to 5.5; Sp. Gr. about 3.5; streak, yellow-brown.

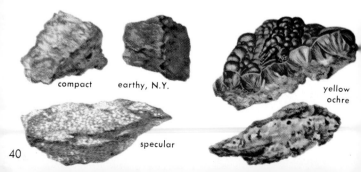

compact   earthy, N.Y.

yellow ochre

specular

IRON ORES

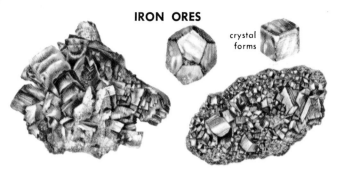

crystal forms

Marcasite (FeS$_2$), sometimes called white pyrite, is similar but lighter and more brittle than true pyrites. The orthorhombic crystals occur in radiating and coxcomb forms, as crustations and concretions, in clays, peat, and coal. H. 6; Sp. Gr. 4.8; crystals common. Specimens commonly crumble and break up on standing.

Pyrite (FeS$_2$) or fool's gold is not like gold at all, but when tarnished may resemble chalcopyrite (p. 32). Used to obtain sulfur and sometimes as a source of iron. Pyrite forms isometric crystals; also occurs as grains or in masses. H. 6; Sp. Gr. about 5; streak, greenish-black; color, light brassy yellow.

Pyrrhotite (Fe$_{1-x}$S) varies in composition but always contains an excess of sulfur. Often found with nickel (Pentlandite) and mined for its nickel content. Occurs as crystals, thin plates, grains, or masses. Color, bronze (pyrite is brassy); H. 4; Sp. Gr. 4.6; streak, gray-black. Often magnetic.

Siderite (FeCO$_3$), is occasionally used as iron ore but deposits are usually small and iron content is low—48 per cent. Crystals common; more often in masses which cleave like calcite (p. 64) Color, yellow, gray, dark brown, H. about 4; Sp. Gr. 3.8; streak, white; luster, pearly.

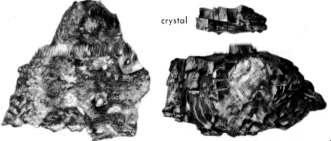

crystal

ore + scrap + limestone + coke + air blast = pig iron

BLAST FURNACE

skip car

hot gases

ore, coke, and lime

hot air blast

slag

iron

Iron has long been man's most important metal. Continuously operating blast furnaces similar to the one shown are used to produce pig iron and ferro-alloys. The furnace is lined with fire brick not bound with mortar.

Iron ore is generally an oxide, as hematite, goethite, or magnetite, although occasionally a sulfide is used after conversion to an oxide by roasting. The ore is dumped into the furnace along with coke, and limestone for flux. Blasts of hot air roar in from the sides, uniting with the carbon of the coke to reduce the ore to metallic iron which flows to the bottom of the furnace, beneath the slag.

The furnaces are tapped regularly. Most of the molten iron is taken directly to Bessemer or open hearth converters to be made into steel. Some is run into molds, making pig iron.

Open-pit mining, Mesabi Range.

U.S. nickel   bar magnet   Canadian nickel

NICKEL ore, hard to smelt in the early days of European mining, was thought bewitched and was spurned by miners. Large deposits are located in Canada, Russia, Australia, New Caledonia, and Venezuela. Niccolite (NiAs) and millerite (NiS) are minor nickel ores occurring with sulfide and arsenic ores of cobalt and iron. The principal ore of nickel (pentlandite) occurs with pyrrhotite (p. 41) and is similar to it. Greenish secondary ores are found near the surface, where weathering has altered the primary minerals. A mine in Oregon produces nickel ores in the U.S. Nickel, a magnetic metal, is widely used in alloys, especially with iron. Permalloy and alnico, both nickel alloys, are used in making magnets. German silver, another alloy, is used for kitchenware, ornaments, and electrical heating wire. U.S. nickels (5-cent pieces) contain about 75 per cent copper and are not magnetic. Canadian nickels, with a higher nickel content, are magnetic.

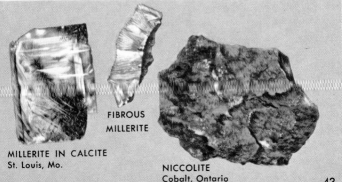

FIBROUS
MILLERITE

MILLERITE IN CALCITE
St. Louis, Mo.

NICCOLITE
Cobalt, Ontario

COBALTITE,
near Salmon, Idaho

0.1 in.

CRYSTAL FORMS: cubic

0.1 in.

12-sided

SKUTTERUDITE,
Skutterud, Norway

COBALT is chemically related to nickel and iron, and its ores are often found with nickel and iron minerals. Democratic Republic of Congo is a large producer; there, the cobalt is found with copper. There are also large deposits of cobalt ores in the Scandinavian countries, Russia, Morocco, and in Canada with nickel.

Cobaltite ($CoAsS$) and skutterudite $(CoNiFe)As_3$ (about the same as smaltite) are the chief ores at Cobalt, Ontario. Both crystallize in the cubic system and look alike. Cobaltite has a hardness of 5.5; color, silvery-metallic. Skutterudite, which is similar, also contains variable amounts of nickel and iron and is classified as an ore of whichever metal is dominant.

Cobalt is a heavy metal (Sp. Gr. 8.9) used in hardening steels and in other alloys. Carboloy, an alloy of cobalt and tungsten carbide, is used in cutting steel. Another ferrocobalt alloy makes permanent magnets. Cobalt oxides are important pigments in paints and ceramics, in which they are used to produce shades of green, blue, yellow, and red. It is the source of color in blue grass. Cobalt also has other chemical uses.

TIN was once used as a coating on the surface of tin cans that protected the iron of the can from rusting. Today most cans are made of aluminum or use other materials as liners. Tin is used in paint and as an abrasive for polishing compounds. Tin is also an important constituent of bearing alloys (Babbitt metal) and for solder. In its oldest use, and still a major one, it is alloyed with copper and some zinc to make bronze.

Tin is essentially a one-ore metal. The ore, cassiterite (SnO_2), tin oxide, contains almost 80 per cent tin. It is commonly brown or black, though occasionally red, gray, or yellow. Streak, pale; luster, glassy to adamantine. Cassiterite forms crystals, but occurs more often as fibrous masses (wood tin) or as crusts or veins in granite and pegmatite rock. It may be distinguished from geothite by its high specific gravity (7.0). The Malay placer deposits of gravel and cassiterite pebbles are mined by huge dredges. Small amounts occur in placer deposits in Alaska, but most tin comes from Malaysia and Bolivia. The Romans mined tin from placer deposits in Cornwall, England.

CASSITERITE

CASSITERITE—WOOD TIN
Guanajuata, Mexico

PLACER CASSITERITE
Malaysia

tin can

typemetal

bearings

solder

SPHALERITE
Joplin, Mo.

crystal 0.2 in.

FRANKLINITE
in calcite
Franklin, N.J.

ZINCITE, WILLEMITE, AND
FRANKLINITE, Franklin, N.J.

ZINC was used as an alloy with copper to produce brass long before it was known as a metallic element. Now zinc is widely used in coating iron to prevent rust (galvanized iron), as well as in dry-cell batteries, paints, other alloys and in chemical industries. Zinc ores occur with lead and copper ores in veins associated with igneous rocks (p. 109) and as replacement deposits in carbonate rocks.

Sphalerite, zincblende, or blackjack (ZnS) is the primary mineral. Color: yellow, brown, or black; luster: resinous. H. 3.5-4. Sp. Gr. 4. Sphalerite has perfect cleavage and breaks easily. Some specimens are fluorescent; others emit flashes of light when scratched in a dark room.

SMITHSONITE ON GALENA
Leadville, Colo.

SMITHSONITE
Broken Hill, New South Wales

HEMIMORPHITE

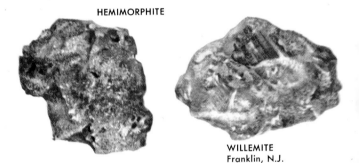

WILLEMITE
Franklin, N.J.

Zincite is orange-red zinc oxide (ZnO), and was an ore at Franklin, N.J. Found with franklinite, a mineral similar to magnetite, containing zinc and manganese.

Smithsonite ($ZnCO_3$) forms as zinc ores weather. It is often a crystalline crust but mostly earthy and dull. Good specimens are found in the Southwest.

Hemimorphite ($Zn_4Si_2O_7(OH)_2 \cdot H_2O$), a zinc silicate with water, forms crystals or earthy deposits.

Willemite (Zn_2SiO_4), translucent, varies in color. Often fluorescent (p. 22).

Tessien, Switzerland

hardness 5

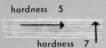

hardness 7

Ivigtut area, Greenland

crystal

emery

Kyanite, or cyanite ($Al_2 SiO_5$), is usually found in schists and gneisses. The white to blue-gray or black crystals are long and blade-like. Luster, glassy to pearly; Sp. Gr. 3.6; hardness unusual—4 to 5 along the crystal axis, but 7 across.

Cryolite (Na_3AlF_6) is a rare mineral. Small amounts have been found in Colorado, but it was the one great deposit in Greenland that was crucial in the history of aluminum, making the smelting of bauxite possible. Now, artificial cryolite is used. H. 2.5; Sp. Gr. 3; glassy or greasy luster. Splinters fuse in candle flame.

Corundum (Al_2O_3) is a primary aluminum mineral found in metamorphosed limestones and in schists. Hexagonal crystals common. Corundum also occurs as dark granules with magnetite—a form known as emery and used as an abrasive. H. 9—harder than any other common mineral; Sp. Gr. 4; color variable.

MINERALS

Bauxite, the ore of aluminum, is a group of related oxides with water of hydration. Most abundant in warmer areas, it forms as aluminum-bearing rocks are weathered. Color, white—though often stained brown or red by iron oxides. H., variable, 1 to 3; Sp. Gr., 2.5. Named after the region near Baux in France where it occurs. Bauxite is found in Arkansas (main U.S. deposit) and in a belt from Alabama into Georgia. The Guianas and Jamaica have rich deposits.

Kaolin, a group of at least three minerals, all aluminum silicates with water ($H_4Al_2Si_2O_9$), is white and scaly when pure. More often found impure as clay; then it is earthy and colored by impurities. Kaolin is widespread, though pure deposits are limited. It is essential in ceramics and has many other uses.

BAUXITE

Arkansas

KAOLIN

CRYSTALLINE KAOLIN
(highly magnified)

49

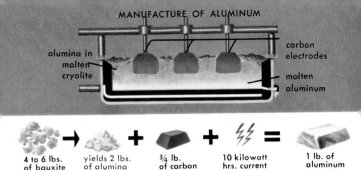

alumina in molten cryolite

carbon electrodes

molten aluminum

4 to 6 lbs. of bauxite → yields 2 lbs. of alumina + ¾ lb. of carbon + 10 kilowatt hrs. current = 1 lb. of aluminum

ALUMINUM, as a commercial metal, has been known for over a hundred years, but its wide use has been much more recent. Clays containing aluminum were used in making pottery long before metals were known. Pottery and other ceramic products still utilize large quantities of aluminum minerals (p. 150). Aluminum is a constituent of feldspars, micas, garnets, corundum, and cryolite, but only the latter has ever been an aluminum ore. Aluminum ores are usually secondary—weathered and altered products of these minerals.

Aluminum, making up over 8 per cent of the earth's crust, is more abundant than iron, but workable deposits are limited. Bauxite, consisting of hydrated oxides of aluminum, is the principal ore. In processing, the 50 to 70 per cent aluminum oxide, called alumina, is first extracted, then dissolved in huge vats of molten cryolite. Alumina is reduced to metallic aluminum by carbon electrodes carrying a strong electric current. It collects at the bottom of the vat. About a million tons a year are produced, mainly for the transportation and construction industries.

Pouring molten aluminum

CHROMITE crystal

CHROMITE—Lancaster, Pa.

CROCOITE—near Dundas, Tasmania

CHROMIUM, a bright silvery metal, has become familiar in the flashy trim on automobiles and household wares. It is often used over a nickel undercoat as a non-rusting plating on iron and steel. Chromium has only one ore, though it occurs in about a dozen minerals.

Chromite ($FeCr_2O_4$), the chromium ore, occurs widely, but workable deposits in the United States are limited largely to California. Commercial deposits occur in South Africa, Cyprus, India, Russia, Brazil, Turkey, and New Caledonia. Chromite occurs in basic igneous rocks or in metamorphic rocks formed from them. Chromite (H. 5.5; Sp. Gr. 4.7) is metallic black or brownish; streak, dark brown. Sometimes slightly magnetic because of its iron content, it occurs in veins or in widespread granular masses, frequently with a coating of serpentine.

Crocoite ($PbCrO_4$), a rare but handsome mineral, is formed when chrome chemicals encounter lead. Attractive crystals of lead chromate then develop.

MANGANESE is widely found, often with iron, barium, cobalt, and zinc. The most common ores are secondary, formed by the action of air and water on manganese silicates and carbonates. Deposits are common in bogs and lakes. Manganese is used to toughen steel for machinery, rails, and armaments. Russia, India, South Africa, and Brazil have most of the ores. U.S. deposits are in Tennessee, Virginia, Arkansas, Arizona, and Montana.

Rhodonite ($MnSiO_3$)—pink, yellow, or brownish—is best known from Franklin, N.J. H. 5.5 to 6.5; Sp. Gr. 3.5; translucent; streak, white. Large, flattened crystals not common. Prismatic cleavage.

Rhodochrosite ($MnCO_3$), softer than rhodonite (H.4), rarely forms large crystals. Translucent with glassy luster; color pinkish, as in rhodonite. Massive, in veins or as crusts. Typical calcite cleavage (rhombohedral).

Romanechite, formerly psilomelane ($Ba_6MN_5O_{10} \cdot 1.3H_2O$), is soft, dull, gray-black, but may form hard, rounded, or stalactitic masses.

Manganite, $MnO(OH)$, is often in prismatic crystals or fibrous masses. H. 4; Sp. Gr. 4.3; streak, red-brown.

Pyrolusite (MnO_2), principal manganese ore, is earthy, powdery, granular, or fibrous. Hardness varies from 1 or 2 up to 6 in crystals. Sp. Gr. 5; streak, black.

MOSS AGATES

India

Montana

California

Manganese oxide dendrites in quartz

RHODONITE—Franklin, N.J.

RHODOCHROSITE
Butte, Mont.

ROMANECHITE
New Mexico

MANGANITE
Ilfeld, Harz, Germany

MANGANESE OXIDE DENDRITES
on dolomite—Princeton, Iowa

PYROLUSITE
Ironwood, Mich.

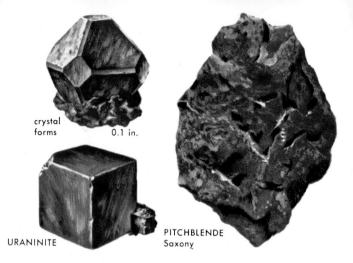

crystal
forms   0.1 in.

URANINITE

PITCHBLENDE
Saxony

URANIUM (discovered in 1789 and isolated as an element about 1842) is now prized as a source of atomic energy. It occurs in more than 50 minerals, most of them rare. The main ores are uraninite and secondary minerals formed from it by weathering. Actually, uranium minerals are widely dispersed in granites and pegmatites. Specimens may be collected wherever these igneous rocks are exposed. Commercial deposits are another matter. Uranium prospecting requires time, patience, and skills which few amateur mineralogists possess.

Radiations disclose uranium.

Uraninite (UO_2) is steel black, opaque, hard (H. 5.5), and heavy; Sp. Gr. 9 to 9.5 for pure specimens. Streak from gray to brown to black; crystals rare. More common is **pitchblende,** a form of uraninite which occurs in massive, fibrous or rounded masses.

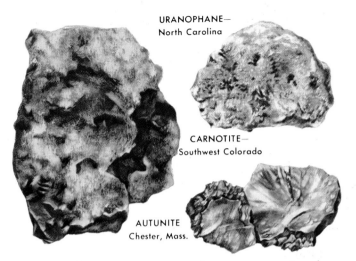

URANOPHANE—
North Carolina

CARNOTITE—
Southwest Colorado

AUTUNITE
Chester, Mass.

Uraninite alters to an orange or red gummy, waxy mineral of variable composition, called gummite.

Carnotite $(K_2(UO_2)_2(VO_4)_2 \cdot 3H_2O)$ is a complex mineral with vanadium and uranium. It occurs in weathered sedimentary rocks as streaks or earthy yellow grains. Common in the Southwest, sometimes on petrified wood.

Uranophane $Ca(UO_2)_2[SiO_3(OH)]_2 \cdot 5H_2O$ is found with uraninite as clusters of tiny yellow, needle-like crystals. It is widely distributed, but never common.

Autunite $(Ca(UO_2)_2(PO_4)_2 \cdot 10\text{-}12H_2O)$ is another secondary uranium mineral. Note the greenish, pearly flecks. Autunite is common in small amounts. This and most other uranium minerals fluoresce strongly under ultraviolet light. Radiation from these minerals can also be detected by the use of a Geiger counter or a scintillometer.

Uranium occurrences

WOLFRAMITE
Luna County, N. Mex.

SCHEELITE
Visalia, Cal.

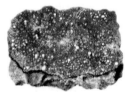

VANADINITE
Tucson, Ariz.

MOLYBDENITE
Climax, Colo.

IMPORTANT MINOR METALS

Wolframite $(Fe,Mn)WO_4$ is a mineral of quartz veins and pegmatites. It is found in the form of tabular or prismatic crystals; dark brown or black with submetallic luster; H. 5.5; Sp. Gr. 7.5; brittle. An ore of tungsten (wolfram)—important in lamp filaments and in steel-alloy cutting tools.

Scheelite $(CaWO_4)$, another ore of tungsten. H. 5; Sp. Gr. 6. Glassy, sometimes transparent; streak, white. Color variable, but light. Found in quartz veins or at contact of igneous rocks and limestone. It is the most important U.S. tungsten ore and it fluoresces blue.

Vanadinite $(Pb_5(VO_4)_3Cl)$ is an attractive, fairly widespread vanadium mineral, but not an important ore. H. 3; Sp. Gr. 7. Vanadinite is a secondary mineral of lead areas. Carnotite (p. 55) is a better source of vanadium, used in alloy steels.

Molybdenite (MoS_2) is a mineral found in pegmatites and veins. It is soft (H. 1.5), metallic and opaque; streak, blue-gray. Occurs as flecks, or tabular crystals. Molybdenum is essential in tool-steel alloys.

Columbite (Fe,Mn)Nb$_2$O$_6$ and **tantalite** (Fe,Mn)Ta$_2$O$_6$ are ores of rare metals. Actually, they are a series of oxides including iron and manganese. The mineral is columbite when the amount of niobium is high, and tantalite when it has more tantalum, which is used as an alloy in surgical instruments. Niobium alloys are used in rocket engines.

COLUMBITE AND TANTALITE
Bedford, N.Y.

The minerals are dark brown or black, crystalline (often twinned). H. 6; Sp. Gr. 5.5 to 8; usually opaque with a submetallic luster. They form in pegmatite, often with tin and tungsten minerals.

Beryl (Be$_3$Al$_2$(SiO$_3$)$_6$) is an ore of beryllium as well as a gem stone (pp. 84–85). The metal is used in alloys of copper and in atomic research. It is lighter than magnesium.

BERYL—Germany

Monazite, H. 5; Sp. Gr. 5, is a complex mineral containing thorium and a number of other "rare earth" metals. It occurs as yellow resinous grains or as larger crystals in pegmatite and in certain sand deposits in southeastern United States. An even greater number of "rare earths" are found in **samarskite,** which occurs rarely in pegmatites.

MONAZITE
Yorktown, N.Y.

SAMARSKITE
Spruce Pine, N.C.

57

Rutile, York Co., Pa.

Gem rutile
(synthetic)

Titanium alloys
used in rockets

Ilmenite, Cumberland, R.I.

Ilmenite crystal

TITANIUM is a metal with a future and its minerals will be of increasing importance. Light weight and a high melting point give it importance in rocket construction. Now used in steel alloys, as a cutting tool (titanium carbide), and in white paints.

Titanium is abundant, making up 0.6 per cent of the earth's crust. Its ores are found principally in southeastern United States and Arkansas, and in India, Norway, France, Switzerland, and Brazil.

Rutile (TiO_2) has varied forms. Most commonly it is black, often in large prismatic crystals. H. 6; Sp. Gr. 4.2; Streak, light brown. Found in igneous and metamorphic rocks. Needle-like crystals are found as inclusions in rutilated quartz. Synthetic rutile is used in rings.

Ilmenite ($FeTiO_3$) is the more common ore of titanium found in many magnetite deposits; associated with gneisses and metamorphic rocks generally. Found as thin sheets, flecks, tabular crystals, grains, or massive. Metallic black in color; opaque. Streak from black to reddish-brown. H. 5 to 6; Sp. Gr. 4.5.

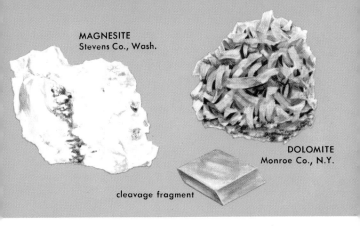

MAGNESITE
Stevens Co., Wash.

DOLOMITE
Monroe Co., N.Y.

cleavage fragment

MAGNESIUM, lighter than aluminum, is the eighth most common element in the earth's crust. It has become a metal of major importance—its alloys find wide use in airplanes, in other sheet metal products, and in casting. Its two principal ores are magnesite and dolomite. Magnesium is also manufactured from sea water—260,000 gallons yielding one ton of the metal. The magnesium chloride is treated with lime from oyster shells. Then the magnesium is extracted by electrolytic action.

Magnesite ($MgCO_3$), usually dull white, massive or granular, sometimes glassy, rarely occurs as crystals. H. 4; Sp. Gr. 3. Large deposits in Washington and California.

Dolomite ($MgCa(CO_3)_2$) is described as a nonmetallic mineral (p. 65), but recently it has come into use as an ore of magnesium—just as the clay minerals may eventually become ores of aluminum. Three light metals—aluminum, magnesium, and lithium—are of growing importance.

Spinel, a magnesium-aluminum oxide, is a well-known gem mineral (p. 85) which is found mainly in Sri Lanka, Burma, and Thailand.

STIBNITE
Shikoku, Japan

STIBNITE, (Sb_2S_3), found with pyrite, galena, and arsenic minerals, is the only common antimony ore. It is more typical of low-temperature veins. Steel-gray, metallic; H. 2; Sp. Gr. 4.5. Crystals common, often found bent. Antimony is used in type metal, pewter, and other alloys. Native antimony is also found as a mineral—but rarely.

ARSENIC is a semi-metal used in alloys. Its poisonous compounds are used in sprays and insecticides.

Realgar (AsS), the simplest ore, forms in low-temperature veins as crusts, grains, flecks, or massive deposits. Crystals are rare. H. 2; Sp. Gr. 3.5; resinous luster; color and streak both orange-red. Realgar slowly breaks down on exposure to light to form orpiment.

Orpiment (As_2S_3), often found with realgar or stibnite, is usually massive, though sometimes sheetlike. Crystals rare. H. 1.5; Sp. Gr. 3.5; perfect cleavage. Specimens become dull on exposure to light.

Arsenopyrite (FeAsS) forms in high-temperature veins and pegmatites. Occurs massive or as crystals. Silvery, metallic; H. 6; Sp. Gr. 6. An ore of arsenic.

ORPIMENT
Manhattan, Nev.

ARSENOPYRITE

REALGAR
Manhattan, Nev.

NONMETALLIC MINERALS

NONMETALLIC MINERALS form a large group of minerals which contain no metal or are not used for the metals they contain. Gems and rock-forming minerals are in separate sections. Even a few metallic minerals (pp. 31-60) belong in the nonmetallic group, for while they contain metals, they are not ores. The nonmetallic minerals are of great importance — for insulation, as fillers, filters, and fluxes, and in the ceramic and chemical industries. The world's most common minerals belong in this group.

SULFUR (S) is a nonmetallic mineral element found in volcanic rocks, around hot springs, and in sedimentary "domes" (p. 145) with salt, gypsum, anhydrite, and limestone. It is yellow (sometimes brown), waxy or resinous, weak and brittle. H. 1.5 to 2.5; Sp. Gr. 2.0. Sulfur's low melting point (110-120°C) aids in mining it from underground deposits. Superheated water is pumped down large pipes, melting the sulfur; compressed air then forces the melted sulfur out. Sulfur is used in papermaking and as a source of sulfuric acid.

SULFUR
on calcite
Texas

crystal

MASSIVE GRAPHITE
New York

pencils

crucible

lubricant

motor brushes

GRAPHITE is one of the world's softest minerals. Diamond is the hardest. Both are carbon (C). Graphite occurs in igneous and metamorphic rocks—schists and marbles. It may form when high temperature veins cut coal deposits, and an artificial form is made in electric furnaces. Graphite is earthy, or forms scaly or flaky crystals with a metallic luster, greasy and flexible. H. 1; Sp. Gr. 2.0. Rich deposits in Korea, Sri Lanka, Mexico, and Madagascar supply the world with natural graphite. Graphite is used for dry and wet lubrication and for electrical and chemical purposes. Its best-known use is for "lead" in lead pencils, where it is usually mixed with other materials to give various degrees of hardness. Graphite is a strategic mineral. Its latest use is as a moderator to slow down neutrons in nuclear reactors.

CALCITE

ICELAND SPAR

TWINNED CRYSTALS

STALACTITES

STALAGMITES

Iceland Spar is a transparent calcite which has the optical property of bending light two ways, making words appear double.

Dogtooth Spar, a common crystalline form, with crystals long and pointed. Crystals from Joplin, Mo., are well known.

Twinned Crystals of calcite, growing together as shown above (left) are very common. This is a typical spectacular form.

Stalactites and Stalagmites form in caves by dripping water. Calcite also forms in thin sheets or draperies from the roof.

DOGTOOTH SPAR

63

BLUE CALCITE
England

TRAVERTINE
Death Valley, Cal.

CALCITE SAND CRYSTALS
Badlands of South Dakota

CALCITE ($CaCO_3$), most common and widespread of the carbonate minerals, is interesting because of its many and varied crystal forms. Calcite occurs in a number of structural forms and frequently grades into dolomite (p. 65). Great masses of calcite occur in limestones (pp. 126–127); small crystal masses are present in rock openings, while rare, transparent crystals occur as Iceland spar, valued for use in optical instruments. Calcite also occurs as a vein mineral in almost all rocks. Crystals are common. H. 3; Sp. Gr. 2.7. Cleavage: perfect, rhombohedral. Most calcite is opaque, slightly colored by impurities; yellow, orange, brown, and green shades occur. Calcite is often fluorescent.

Other Forms of Calcite

Nail-head Spar: flattened, rhombohedral crystals, often in clusters.

Travertine: a general term for massive, noncrystalline calcite as found in caves (also called cave onyx). Opaque, often colored.

Tufa: porous, white travertine from spring deposits.

Chalk: white, soft, compact shells of small sea animals.

ARAGONITE TWINNED CRYSTALS
Sicily

PRECIOUS CORAL
Italy

ARAGONITE CRYSTAL
Cambria, England

ARAGONITE is chemically the same as calcite ($CaCO_3$) but is less common and crystallizes in the orthorhombic system. It is slightly harder and heavier—H. 3.5 to 4; Sp. Gr. 2.9. Aragonite does not cleave as distinctly as calcite, although it, too, bubbles strongly in dilute hydrochloric acid. Aragonite is usually white, gray, or cream. The mother-of-pearl lining of sea shells is aragonite. **Flos ferri** is a branching growth of pure white aragonite in mines and caves. **Coral,** formed by plants and animals in warm seas, is commonly aragonite. Precious coral is valued for gem and ornamental use.

DOLOMITE ($MgCa(CO_3)_2$) occurs in large bedded deposits and as veins in other sediments. There are extensive deposits in the Austrian Tyrol—the Dolomite Alps. Dolomite is both a rock and a mineral (p. 109). The best mineral specimens come from veins or limestone cavities, and include crystals with curved faces. Dolomite is harder than calcite (H. 3.5 to 5; Sp. Gr. 2.8), but similar in crystal form and cleavage. Reacts slowly with hydrochloric acid. White or varicolored. Widely distributed.

DOLOMITE CRYSTALS
Mexico

crystal
Trumbull County, Ohio

fishtail twins

curved crystal

GYPSUM is a common mineral. It forms a sedimentary rock precipitated from evaporating sea water under dry or arid conditions. Massive beds of gypsum are found in over a dozen states and in Maritime Canada. When mined, gypsum is the basis of a large industry producing paints, plaster, plasterboard, tile, and other construction materials. A small amount of gypsum in Portland cement keeps it from setting too fast.

Gypsum was known and used in Europe for centuries. It was first burned in open fires, later in kilns. Heating gypsum drives out part of the water; the burned gypsum, ground to a white powder, is known as plaster of Paris, as it was first made near there. When moistened, plaster of Paris absorbs water again and hardens as gypsum rock, so it is used in making plaster casts and in plastering and as an ingredient of many prepared construction materials.

Crushed gypsum is used in agriculture, being added to the soil as "land plaster." It aids the growth of peas and neutralizes alkaline soils. Over 19 million tons are mined annually in this country—about 18 per cent of the world production.

Gypsum is a colorless or white mineral, sometimes tinted by iron or other impurities. It is calcium sulfate, with water ($CaSO_4 \cdot 2H_2O$). H. 2, Sp. Gr. 2.3. Luster is pearly, glassy—sometimes fibrous. Streak, white. Its crystals have two cleavage planes—one perfect. Does not bubble in cold hydrochloric acid, but will dissolve in hot.

Most gypsum occurs in bedded deposits. A compact massive form known as alabaster is carved for ornaments. Crystalline gypsum (selenite) occurs in caves and limestone cavities—also in clays, shales, and some sands. Crystals may grow several feet long, and may be twinned or curved.

Anhydrite ($CaSO_4$) is chemically similar to gypsum, but does not contain water. It is found in crystalline masses, though good crystals are rare. Sometimes it is fibrous, granular, or scaly. Color white to gray; streak, white. Luster, glassy or pearly, often translucent, rarely transparent. H. 3 to 3.5; Gr. 2.9. It is often found with gypsum, and may change to gypsum by absorbing water. Also occurs in salt beds and with sulfur in "domes."

MASSIVE GYPSUM
Grand Rapids, Mich.

FIBROUS GYPSUM
Mammoth Cave, Ky.

ANHYDRITE
Mound Prairie, Minn.

crystal

HALITE (rock salt)
Detroit, Mich.

crystal forms

cleavage fragment

HALITE or common salt is sodium chloride (NaCl). It has been used since prehistoric days, and there is no substitute for it in nutrition or in industry. All halite comes from the sea. Layers of rock salt mark areas where seas dried up in ancient times. In many places salt is still made by evaporating sea water in shallow basins.

Halite is colorless when pure, but is usually discolored some shade of yellow, blue, red, gray, or brown. It is transparent to translucent, brittle, and with excellent cleavage parallel to its crystal faces. H. 2 to 2.5; Sp. Gr. 2.3. It occurs in granular, fibrous, or crystalline masses, easily recognized by the cubic crystals and by the mineral's familiar taste. Halite is rarely pure. It occurs with other salts of calcium and magnesium.

GLAUBERITE, a sulfate of sodium and calcium, occurs around mineral springs and with other evaporites. H. 2.5; white to gray.

SALTPETER, a fertilizer, is sodium nitrate ($NaNO_3$). It occurs in Chile and our Southwest—with gypsum, halite, and glauberite.

California

Chile

Michigan salt mine.

Salt and Related Minerals are part of the "alkali" which makes some western soils difficult to use for agriculture. These minerals form under semi-arid to arid conditions, often with borax (p. 70). Halite is mined by shaft mining or by pumping water into the deposit and later pumping out the brine. In purification, potassium and magnesium salts, bromine, and iodine are obtained as by-products. The halite is recrystallized, becoming very pure in the process. In addition to its use in food and as a preservative, salt is essential in chemical industries, in the manufacture of soda ash for glass products, and in soapmaking and metallurgy. Chlorine from salt is used as a bleach and in water purification.

"Twenty-mule team" hauling borax—Death Valley, Cal.

BORAX, originally obtained from deposits of volcanic origin, now comes mostly from brines and dry lake beds where ground water has concentrated the borax. Used in making glass and enamels and in chemical industries.

Borax, sodium borate with eight parts of water, occurs in alkali lakes or as a crust on the desert soil. Once hauled in twenty-mule teams from the Death Valley region. White to gray; H. 2, Sp. Gr. 2.7; glassy.

Kernite is chemically similar to borax but contains only three parts of water. It is often colorless and transparent. H. 2.5; Sp. Gr. 2.0. Streak white, luster glassy.

Colemanite is a calcium borate with five parts of water. Prismatic crystals common. Color, white. H. 4 to 4.5; Sp. Gr. 2.3. Streak, white. Luster, glassy to dull. Often found in white, chalky or hard, glassy masses.

BORAX
Esmeralda, Nev.

KERNITE
Kramer, Cal.

COLEMANITE
southern California

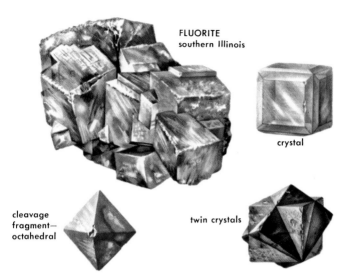

FLUORITE
southern Illinois

crystal

cleavage
fragment—
octahedral

twin crystals

FLUORITE in commercial quantities occurs in both sedimentary and igneous rocks. Veins of fluorite (CaF_2) with quartz or calcite sometimes contain lead, copper, and zinc minerals. Mexico, China, and Russia are the leading producers of fluorite. Fluorite is used to produce a fluid slag in steel-making and in smelting ores. It is used in making high-test gasoline, Freon, and many other chemical products. Fluorite is a most attractive mineral of varied colors—white, blue, green, and violet. Transparent to translucent. H. 4; Sp. Gr. 3.2. Streak, white; glassy luster. Good octahedral cleavage. Crystals and fine cleave fragments make attractive specimens for the collector. Often fluorescent in ultraviolet light.

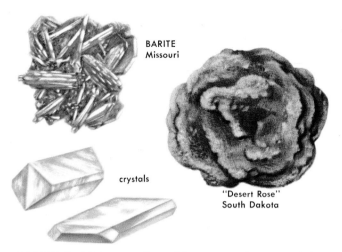

BARITE
Missouri

crystals

"Desert Rose"
South Dakota

BARITE ($BaSO_4$) is a sulfate of barium, a silvery metal. It is placed with the nonmetallics because it is rarely a source of the metal and is widely used otherwise. Barite occurs in many ways, occasionally as large transparent or translucent crystals; sometimes as crystalline vein filings, with fluorite or calcite, or with metallic ores. Large concretions are found in South Dakota and smaller rose-shaped ones ("desert roses") in the sands of Oklahoma.

Barite is a fairly soft (H. 2.5 to 3.5) but dense (Sp. Gr. 4.5) mineral, colorless to white, yellow, gray, and brown. Luster, glassy, sometimes pearly; streak, white. Sometimes granular or earthy. Crystals are common, often broad and thick. Barite is used in making lithopone for paint, as an aid in well drilling, in making glass, as a filler in glossy paper, and in ceramics.

MASSIVE APATITE
Norway

APATITE CRYSTALS
in calcite
Ontario, Canada

APATITE gets its odd name from the Greek word meaning "to deceive" because its varied forms and colors caused early mineralogists to confuse it with a half-dozen other minerals. Apatite may be transparent, translucent, or opaque, with a color that varies from white to brown, green, yellow, or violet. Apatite occurs in veins with quartz, feldspar, and a variety of iron ores. Hexagonal crystals are common, some gemmy. Apatite is a phosphate of calcium, usually with some fluorine ($Ca_5(Cl,F)(PO_4)_3$). H. 5; Sp. Gr. 3.2; luster, glassy; streak, white. Outstanding commercial deposits are present in Ontario and Quebec, Canada and Mexico. Phosphate rock, used for fertilizer, contains minerals related to apatite. A thin chip of apatite will color a gas flame orange (1). When wet with sulfuric acid *(caution!)* it colors the flame a pale bluish-green (2) due to liberated phosphorus.

FOLIATED TALC
St. Lawrence County, N.Y.

GRANULAR TALC
St. Lawrence County, N.Y.

STEATITE
Lancaster, Pa.

TALC may form when magnesium-rich rocks are altered, especially by heated waters. Hence talc occurs as a secondary mineral with serpentine (p. 107), chlorite (p.106), schists, and dolomite. It is found in irregular deposits in metamorphic rocks. Talc has a variety of uses ranging from cosmetics (talcum powder) to fillers in paint, insecticides, rubber, and paper. The use depends on the form in which the talc occurs—this may be massive or fibrous.

Soapstone is rock usually rich in talc. **Steatite** is a massive talc, usually of high grade. Talc is frequently sheetlike (foliated) or granular. As a mineral, it is one of the softest (H. 1 to 1.5), Sp. Gr. 2.7. Foliated talc has good basal cleavage and breaks somewhat like mica, though the scales are not elastic. Translucent to opaque; color varies, from white to greenish, yellow, or pink. Pure talc has a greasy, soapy feel and when powdered acts as a lubricant.

Talc and soapstone were known in ancient times. Eskimos carved lamps and pots from it, as did the Mound Builders. The Egyptians, Babylonians, and Chinese also made use of it.

ASBESTOS is the name given to a group of minerals different in origin but of similar appearance. Originally the term was applied to fibrous minerals closely related to amphibole (p. 100). Of the fibrous amphiboles, **crocidolite** or blue asbestos is best known. Mountain leather is a heavy, matted form; its fibers are usually short and brittle.

CROCIDOLITE ASBESTOS

The best-known source of asbestos is serpentine (p. 107) known as **chrysotile.** It is silky, fibrous, and strong. Most chrysotile comes from the famous Quebec and Vermont deposits, where it honeycombs the serpentine rock. Here the cross fibers vary from ½ to 3 inches in length.

CHRYSOTILE ASBESTOS
Quebec, Can.

Chrysotile fibers are so fine they can be divided into almost invisible strands. They spin well and were used in woven insulation and for fireproofing. The fibers do not burn and they conduct heat very slowly. Shorter fibers were mixed with gypsum to make asbestos board. In recent years health concerns related to exposure to asbestos fibers have severely limited its use. Sight identification of chrysotile is easy—no other mineral is this fibrous.

LONG-FIBERED CHRYSOTILE
Thetford, Quebec, Can.

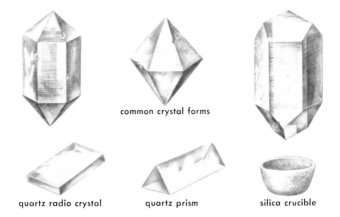

common crystal forms

quartz radio crystal   quartz prism   silica crucible

QUARTZ is one of the most common minerals in the earth's crust. As the chemical silica (SiO_2) it forms an important part of most igneous rocks. Some sandstones are almost 100 per cent quartz, and so are such metamorphic rocks as quartzite. At different temperatures quartz crystallizes in different ways. One variety, alpha-quartz, changes to another, beta-quartz, at 573°C, and changes back when cooled.

When conditions permit, quartz forms hexagonal crystals (pp. 78–79). Doubly terminated quartz crystals (often found in limestone cavities) have such local names as Herkimer diamonds. Larger crystals are found lining cavities; they are often cut as gem stones and are sold as rock crystal. Arkansas and Brazil are famous for its deposits of crystal quartz. Quartz also occurs in a form in which the crystals are of microscopic size and hence not apparent (pp. 80–81). Silica combined with water is opal (p. 82).

Rock crystal on drusy quartz (crust of small crystals).

Quartz is sometimes colorless but more commonly white, sometimes yellow, brown, pink, green, blue, or black. Its luster is glassy in crystalline forms, waxy or greasy in chalcedony. H. 7; Sp. Gr. 2.6; streak, white; brittle with conchoidal fracture. Quartz is best recognized by its hardness, luster, and occurrence, and by its crystal form when evident. Many kinds of quartz are valued as gems and are classed as semiprecious stones.

When crystal quartz is cut at a particular angle to its axis, pressure on it generates a minute electrical charge. This effect makes quartz of great usefulness in the electronics industry and clocks. The supply of natural high quality quartz is so limited that methods for growing quartz crystals in the laboratory have been developed. Quartz transmits short light waves (ultraviolet) better than glass. When not of high quality, crystal quartz is melted to form blanks to make special lenses and prisms. Optical quartz can be made only from crystals. Quartz sand is used in making glass.

CRYSTALLINE QUARTZ is the most common kind, though well-developed clear crystals are relatively rare. Rock crystal (colorless crystal quartz) makes a fine gem. The colors in crystal quartz may in part be due to impurities (aluminum and iron) and to radioactivity, as in smoky quartz; some colors disappear when the quartz is heated. The crystals often include air bubbles and traces of other minerals. Hairlike crystals of rutile form rutilated quartz. Cat's eye and tiger's eye may contain fibers of asbestos. Ferruginous quartz is colored by hematite.

Rose Quartz occurs in crystalline masses—rarely as individual crystals. Some rose quartz is asteriated—that is, the cut stone reflects or transmits light in a starlike pattern. The pink color is due to inclusions of tiny crystals of the mineral dumortierite.

Blue Quartz is an uncommon variety, different from the violet amethyst. It is found in the Blue Ridge Mountains, and with blue feldspar in the Smokies.

Citrine is a yellow quartz also called false topaz because of its color. Good crystals of gem quality come from Brazil. Not to be confused with pale smoky quartz.

Amethyst color is due to traces of iron. Deeper colored specimens are cut as gems. Once highly prized, amethyst lost much of its value after great Brazilian deposits were found.

Smoky Quartz, also called cairngorm or Scotch topaz, varies in color from smoky yellow to brown and black. The latter form is called morion. A well-known Scotch gem stone, though also found and prized elsewhere.

Milky Quartz, found in veins, is the most common crystalline quartz. It is translucent to opaque and is not often used as a gem.

QUARTZ WITH INCLUSIONS

| tiger's eye | with rutile | with iron oxide |

78

ROSE QUARTZ
South Dakota

AMETHYST
Brazil

SMOKY QUARTZ
Scotland

BLUE QUARTZ
Pennsylvania

CAIRNGORM (MORION)

CITRINE QUARTZ
Brazil

MILKY QUARTZ
New Hampshire

CRYPTOCRYSTALLINE QUARTZ merely means quartz with microscopic crystals, in contrast to the varieties described on the preceding page. This group includes the chalcedonies and the flints, cherts, and jaspers. Most of these are translucent or opaque; some are colorful and are prized as gems.

Chalcedony is a group term for a waxy, smooth form of quartz often lining cavities, filling cracks or forming crusts. Sometimes transparent, usually translucent. Colors from white to gray, blue, brown, or black.

Carnelian (sard) is a translucent chalcedony, usually some shade of red or reddish brown.

Jasper is an opaque quartz usually red, yellow, or brown, or a mixture of these colors. Sometimes banded. May grade into chert.

Flint is a gray, brown, or black quartz frequently found as nodules in chalk. Duller, more opaque and rougher than chalcedony, it breaks with conchoidal fracture, producing sharp edges. Widely used by early humans for making tools.

Chrysoprase is a translucent, apple-green chalcedony; coloring due to nickel.

Agate is chalcedony with a banded or irregular, variegated appearance. Bands may be wavy or parallel, from differences in deposition. Petrified wood is usually an agatized wood. Agate may be artificially colored. See p. 52 for moss agate.

Onyx is agate with even, parallel bands usually of black and white or brown and white.

Sardonyx is a form of onyx with alternating bands of sard (carnelian) and white—that is, of red and white bands.

Chert, or hornstone, is an impure form of flint—usually more brittle. Color: white, yellow, gray, or brown.

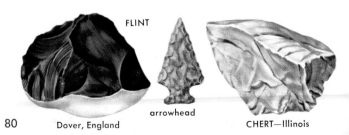

FLINT

arrowhead

Dover, England

CHERT—Illinois

CARNELIAN—Brazil

CHRYSOPRASE—California

ONYX—Brazil

BANDED AGATE—Mexico

EYE AGATE—Brazil

CHALCEDONY—Tampa, Fla.

JASPER—Michigan

81

COMMON OPAL
Sonoma Co., Cal.

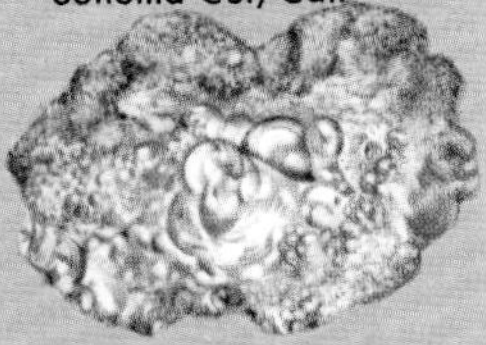

HYALITE
Waltsch, Bohemia

GEYSERITE—Wyoming

DIATOMS (magnified)

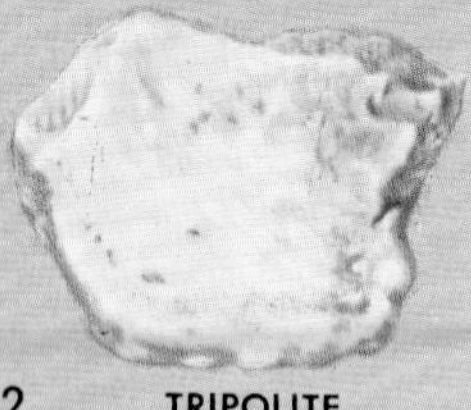

OPAL is a noncrystalline form of silica containing varying amounts of water (usually 3 to 9 per cent). Opal forms as a low-temperature deposit around hot springs and in veins, as in Nevada. It makes up the skeletons of diatoms and siliceous sponges. Opal is usually colorless or white, with color only in gem forms. H. 5.5 to 6.5; Sp. Gr. 2.1. Luster glassy, pearly, or resinous. Streak white. Among the varieties are:

Common Opal Milky white, green yellow to brick red; somewhat translucent, glassy, or resinous. Widespread, often in volcanic rocks. Various names for different colors and gem forms (p. 86).

Hyalite is a clear, colorless, bubble-like, glassy opal encrusting rocks or filling small veins. Sometimes translucent or white.

Geyserite or siliceous sinter is a form of opal found around geysers and hot springs. It may be firm, porous, or fibrous; usually gray or white, and opaque. Often forms odd structures.

Tripolite is formed of microscopic shells of diatoms (diatomite) and other organisms. White, chalky, fine-grained, but hard. Will scratch glass.

TRIPOLITE
Massachusetts

GEM MINERALS

GEMS are the most prized and famous of all minerals. All are better, clearer, or more colorful forms of minerals which in common occurrences are less beautiful and less spectacular. Diamonds, emeralds, rubies, and sapphires stand out as the most highly valued gems. Other stones are commonly referred to as semiprecious and ornamental stones. Scarcity and fashion are important in determining the value of a gem, but the following physical properties are prized: luster, transparency, color, and hardness.

Luster depends on how light is reflected by the mineral. The transparent gems also refract or bend light and are cut to turn the light back into the observer's eye. Color is essential in some gems, and incidental in others. It may add or detract greatly from the gem's value. The harder the gem, the better it resists scratching of its polished surfaces. Here are some of the best-known gems and semiprecious stones. Many other minerals and some rocks are occasionally used as gems

brown

yellow

green

DIAMONDS are pure carbon (C); H. 10; Sp. Gr. 3.5. Found as isometric crystals or crystalline masses. Colorless or with tints of yellow, pink, blue, brown, and black. When not of gem quality, diamonds have important industrial uses as abrasives.

DIAMOND IN MATRIX
South Africa

TRANSPARENT GEMS are striking for their luster and brilliance and often for hardness and color too. Most are oxides of aluminum, beryllium, and magnesium, sometimes with silica. All quartz gems (pp. 86–87) are silica.

AQUAMARINE CRYSTAL
North Carolina

Aquamarine is a light blue-green variety of beryl ($Be_3Al_4Si_6O_{18}$), an ore of the metal beryllium. H. 7.5 to 8. It also occurs as yellowish golden beryl. Found in Brazil, Pakistan, Russia, China, Nigeria, and in many New England localities.

CHRYSOBERYL—New York

Chrysoberyl differs from beryl in being $BeAl_2O_4$. Hardness 8.5; Sp. Gr. 3.6. Color yellow due to iron. Alexandrite is green in sunlight, red in incandescent light.

EMERALD—North Carolina

Emerald is a variety of beryl varying in color from light to deep emerald green. Most come from Colombia.

RUBY
CRYSTAL
Sri Lanka

Corundum gems are all rare forms of alumina (Al_2O_3). The gems vary in color. Deep red rubies are valued more than diamonds. Star sapphires reflect light in a six-pointed star, as do a few other minerals. Important sources are Myanmar and Sri Lanka.

"brilliant"
cut ruby

STAR SAPPHIRE
Tasmania

cut sapphire

TOPAZ, a mineral of granites and other igneous rocks, is an alumino-fluoro-silicate. Large crystals have been found, some of gem quality. These are usually yellow, brown, or pink (when heated). H. 8; Sp. Gr. 3.5. False topaz is brownish quartz.

GARNETS are a common group of silicate minerals (pp. 104–105), sometimes of gem quality. Pyrope and almandine are best known. Green demantoid is also a gem.

SPODUMENE (LiAlSi$_2$O$_6$) occurs in pegmatite rocks, often forming long crystals. Two gem forms exist: Hiddenite, a green gem spodumene from North Carolina, and kunzite, pink colored, first found near San Diego, Cal. Gem spodumenes also occur in Brazil and Afghanistan.

ZIRCON (ZrSiO$_4$) is common in igneous rocks, but fairly rare as a gem stone. Clear brown crystals turn blue when heated and hence make better gems.

TOURMALINE, commonly black (p. 95), forms long crystals, sometimes varicolored. Red, green, brown, and blue tourmalines are known.

SPINEL (MgAl$_2$O$_4$) sometimes reaches gem quality, the best red gems coming from Sri Lanka. Brown, green, and even blue spinels occur. H. 8; Sp. Gr. 3.8.

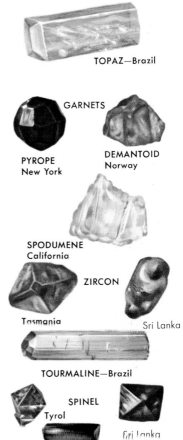

TOPAZ—Brazil

GARNETS

PYROPE
New York

DEMANTOID
Norway

SPODUMENE
California

ZIRCON

Tasmania

Sri Lanka

TOURMALINE—Brazil

SPINEL
Tyrol

Sri Lanka

Some transparent gems are identical minerals that differ only in color, as the ruby and the sapphire. Transparency, lack of flaws, color, and size determine the value of these gems. For synthetic gems, see pp. 92–93.

85

QUARTZ GEMS, the best known common gems, are the same as the minerals described on pp. 76–82. Of these gems, opals are the most valuable, some being classified as precious. The transparent quartz gems range from colorless through yellow, brown, blue, black, purple, pink and,

ASTERIATED ROSE QUARTZ

CITRINE

SMOKY QUARTZ

AMETHYST

THREE CUTS OF ROCK CRYSTAL
Brazil

PRECIOUS OPAL
Australia

FIRE OPAL
Mexico

BLACK OPAL
Australia

rarely, green. The translucent or opaque quartz gems have an even wider array of colors and forms. Some are banded, striped, or mottled. Names of all these gems vary locally. Some bear several names; some names are used for several stones.

WHITE CHALCEDONY—Uruguay

AGATE—Iowa

SARD

ONYX—Brazil

JASPER—Montana

BLOODSTONE

CHRYSOPRASE—Cal.

CARNELIAN—N. Mex.

Myanmar

Australia

CARVED JADE
early Chinese

JADE is the name given a group of opaque, waxy or pearly minerals, usually green but also yellow, white, or pink. There are two kinds of "true jade." One is jadeite, a gem form of pyroxene. The other is nephrite, a form of amphibole (pp. 100–101). Light, translucent, emerald green jadeite is considered a precious stone.

MOONSTONE is albite (p. 99) with a bluish sheen.

AMAZONITE is a green variety of gem-quality microcline (p. 99).

LAPIS-LAZULI is a rock rich in lazurite, and sometimes with pyrite. Usually an ornamental stone.

MALACHITE, often with azurite (pp. 32–33), occurs in masses. Cut for gems or ornaments.

OPAQUE GEMS, with the exception of jade, grade off into ornamental stones. The group includes representatives of metallic ores and rock-forming minerals. Some, like obsidian, lapis, and jet, are better classified as rocks. One unusual mineral that could fit in this group is the

RHODONITE may be used as a gem stone because of its color and hardness. See also pp. 52–53.

RHODOCHROSITE, softer than rhodonite, has the same attractive pink color. See also pp. 52–53.

HEMATITE, an iron mineral (pp. 40–42), is cut as a black, shiny gem when crystalline.

JET, a tough form of lignite coal, takes a high polish. A semiprecious gem from England and Spain.

OBSIDIAN or volcanic glass (p. 116) has long been used for arrowheads and primitive cutting tools. It polishes well and makes an attractive semiprecious stone.

TURQUOISE, a copper-and-aluminum phosphate mineral, is prized by western American Indians. Good quality turquoise is rare. Cheap stones are often dyed blue.

pearl, formed by a number of freshwater and marine mollusks when sand or some other material irritates the animal's mantle. Layers of aragonite form the pearl, which grows year by year. Pearls are soft (H. 4), but have a unique luster.

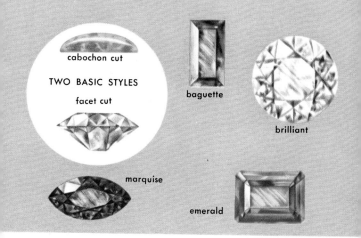

TWO BASIC STYLES

cabochon cut

facet cut

baguette

brilliant

marquise

emerald

GEM CUTTING is an ancient art which has often been successfully pursued by amateurs. With comparatively little equipment a person can cut and polish gems and ornamental stones and can learn to mount them in hand-made or purchased settings. As a creative, satisfying hobby, gem cutting grows more popular each year. Enthusiasts often form clubs for mutual aid or do their work as one activity of a rock and mineral club. Skill in cutting and polishing comes with practice. If one cannot find gems or semiprecious stones, rough material can be purchased from dealers by the piece or by the pound or carat, depending on the stone.

With apparatus powered by a small electric motor, the amateur first learns to cut stones and make cabochons,

Dop stick holds gems for cutting and polishing

simple gem cutting
and polishing outfit

pieces with round or curved surfaces. Later he learns to cut stones with facets or faces, which brings out the brilliance of transparent gems. Five common types of faceted cuts are shown at the top of p. 90.

As in most other hobbies involving skill, one learns best by working and studying with a person of experience. After you have begun, the books and magazines on gems and gem cutting listed below will be of value.

Sinkankas, John, **Gem Cutting: A Lapidary's Manual,** 3rd edition, Van Nostrand and Co., New York, 1984. A very good introduction to faceting gemstones by one of the best faceters in America.

Hall, Cally, **Eyewitness Handbook to Gemstones,** D. K. Publishing, Inc., New York, 1994. A very good introduction to gemstones and where they are found.

The Lapidary Journal, Circulation Department, P.O. Box 56289, Boulder, CO 80323-6289. Published bimonthly, this magazine has many articles devoted to the fashioning of minerals into art objects.

diamond wheel
cutting agate

diamond wheel

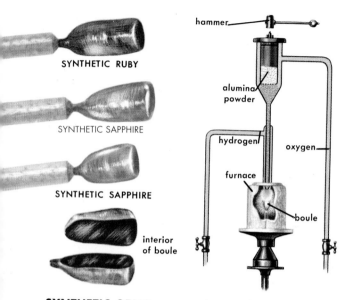

SYNTHETIC RUBY

SYNTHETIC SAPPHIRE

SYNTHETIC SAPPHIRE

interior
of boule

hammer

alumina
powder

hydrogen

oxygen

furnace

boule

SYNTHETIC GEMS—once a dream—during the past few decades have become a reality. Besides the economic problems there were ample scientific ones. Attempts to make synthetic diamonds about the turn of the century met with dubious success. The problem was solved by renewed efforts in 1955. The goal was to produce industrial diamonds, rather than gems. It is with corundum, however, that the best synthetic gems have been made. A method of fusing fine alumina (Al_2O_3) in a very hot flame was perfected in 1902. By adding the appropriate mineral pigments, synthetic rubies and sapphires of large size and fine quality have been formed.

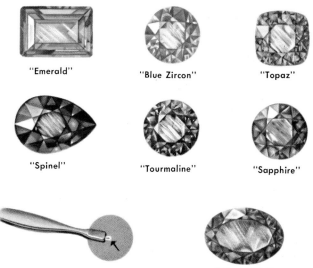

"Emerald" "Blue Zircon" "Topaz"

"Spinel" "Tourmaline" "Sapphire"

SYNTHETIC DIAMOND (actual size) "Alexandrite"

In the apparatus shown on p. 92, the alumina mixture sifts down through the oxyhydrogen flame and forms a slow-growing *boule* at the end of the ceramic rod. Synthetic jewels are universally used for watches, and it is hard to tell synthetic corundum gems from natural ones.

Besides the corundum gems, a number of other synthetic gems have been produced. These are chemically identical to the natural gems. Other man-made gems are without natural counterparts and, finally, there are imitations made of glass. Gems may also be dyed (as agates) or have their color changed by heat, chemical action, or radioactivity. A thin layer of precious stone is sometimes mounted on a larger backing of inexpensive material (p. 15). For these and other reasons it is wise to get expert help when selecting valuable stones.

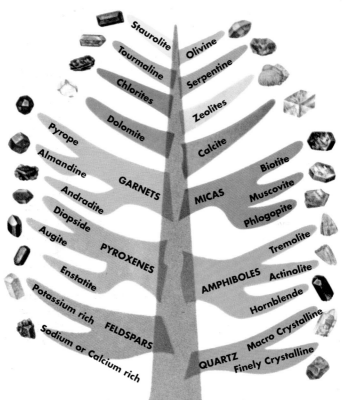

ROCK-FORMING MINERALS

The minerals on this tree are the ones that are of major importance in forming the rocks of the earth's crust. Some rocks have no definite minerals—just organic or glassy material. But most of them do contain discrete minerals—those from the groups pictured above predominating.

ROCK-FORMING MINERALS

ROCK-FORMING MINERALS are important as the building blocks of the solid earth, from which mountains are made and valleys carved. They furnish the minerals of our soil and the salt of the seas. Most of these minerals contain metals, but they are not metallic ores. Some are valued as gems when, under rare circumstances, they gain gem quality. A few are of commercial value, but it is as basic constituents of rocks (p. 109) that the true value of the group is realized.

Nearly all the rock-forming minerals are silicates, that is, they consist of a metal combined with silicon and oxygen. Some are complex silicates, involving several metals and several silicate groups. This complexity makes chemical testing of rock-forming minerals a difficult matter for the amateur mineralogist. But the common minerals of igneous and metamorphic rocks can be identified by paying close attention to their physical properties.

In the rock-forming group are some minerals treated elsewhere. Quartz (pp. 76–82), especially in its crystalline forms, is very common in rocks. Calcite and dolomite (pp. 63–65) can also be called rock-forming minerals; so can halite and gypsum.

BLACK TOURMALINE
Tyrol

TOURMALINE, a silicate of aluminum with boron and several other metals, is occasionally abundant with mica and feldspars in granitic rocks. It is mostly black—other colors forming gems (p. 85). Note the triangular, striated crystals. H. 7; Sp. Gr. 3.

MICAS are an unusual family of minerals, famed because of the perfect basal cleavage which enables one to cleave off paper-thin, flexible sheets. Such sheets from large "books" made the "isinglass" heatproof windows of old stoves and ranges. Because of their high electrical resistance, the iron-free micas are widely used in many kinds of electrical and electronic equipment. This important use has led to the experimental production of artificial micas.

Micas are silicate minerals. All include oxides of aluminum and silicon with other metals, singly or in combination. They also contain some water in combination with the other elements. Part of this water is lost when micas are heated. Micas are common in granites and similar igneous rocks. Large, six-sided crystals—some weighing as much as 100 lbs.—occur in pegmatites. Muscovite, the most common mica, is mined commercially in the Black Hills of South Dakota and in a few other places in this country. The best and most perfect "books" are from large deposits in India.

Micas also form in metamorphic rocks as other minerals are altered by heat and pressure. The mica in mica schist and in gneiss is of this origin, as is the mica in some kinds of crystalline marble.

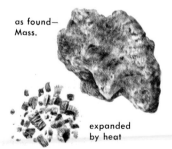

as found—
Mass.

expanded
by heat

VERMICULITES are clay minerals containing water, closely related to the micas. When they are heated, steam forms and practically explodes the flaky mineral, expanding it to many times its original volume. A soft yellow or bronze material, it is used for insulation and for lightweight aggregates. It is excellent for growing cuttings and seedlings.

BIOTITE is a dark-colored mica, brown or black, sometimes green, containing magnesium and iron. It is abundant in some granites and is also common in schists and gneiss. Small barrel-shaped crystals are sometimes found. Biotite may occur with muscovite in metamorphic rocks. Thin cleavage sheets often show light spots, rings, or halos. H. 2.5–3; Sp. Gr. 2.9.

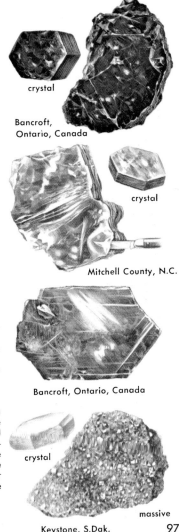

crystal

Bancroft,
Ontario, Canada

MUSCOVITE is usually a pale, almost colorless mica—H. 2.2; Sp. Gr. 2.8. It is a potassium mica of variable chemical composition. Named after Muscovy, where it was used as a substitute for glass, this common mica occurs in many places. Crystals are common and may include flattened garnets, quartz, or tourmaline. Abundant in granites and pegmatites.

crystal

Mitchell County, N.C.

PHLOGOPITE, related to biotite and found with it, is a magnesium-potassium mica often containing iron and fluorine. Hardness and specific gravity much like muscovite. Large crystals of phlogopite are mined in Ontario and Madagascar. It is the mica usually seen as brown flecks in crystalline dolomite and marble.

Bancroft, Ontario, Canada

LEPIDOLITE is a lithium mica with potassium and fluorine, also quite variable in composition. It is an ore of the light metal lithium. Some deposits occur in New England; more near San Diego, Cal., where one finds both an attractive lavender and a pale yellow form. Lepidolite gives a crimson flame color.

crystal

massive

Keystone, S.Dak.

97

LABRADORITE — an iridescent blue plagioclase feldspar used in ornament and decoration.

FELDSPARS form the most abundant group of minerals. If the group were considered a single mineral, it would be the most common mineral by far—five times as common as quartz. Feldspars are found in nearly all igneous rocks and in rocks formed from them. All are aluminum silicates combined with one or two more metals. Feldspars have common physical properties. Their crystal forms are very similar and the crystal angles are all close to 60° and 120°. Feldspars show two good cleavage faces, at right angles or nearly so. Their hardness is 6 or a bit more, and their specific gravity about 2.6. They usually have a smooth, glassy, or pearly luster.

Feldspars may be classified by crystal structure or by chemical composition. Neither is of much help to the amateur. From the chemical point of view, the potassium feldspars (orthoclase and microcline) are put in one group. The other group, containing the plagioclase feldspars, begins with albite (a sodium feldspar) and ends with anorthite (a calcium feldspar). Between these are feldspars that are difficult to identify because they contain varying proportions of sodium and calcium, as oligoclase, andesine, and labradorite. Rarer feldspars with barium and other metals are also known to occur.

Feldspars are widely used in ceramic industries in the manufacture of glazes, flux enamels, and binders. They ultimately weather to form kaolin (p. 49) or other clay minerals (p. 150).

ORTHOCLASE, a fairly common feldspar, is usually white, yellow, or pink. It is a potassium feldspar, some varieties containing sodium; in others barium may replace the potassium. A variety known as adularia, with a bluish reflected sheen, is sometimes cut as moonstone, a semiprecious stone. Another variety, sanidine, is found in volcanic rock.

MICROCLINE is also a potassium feldspar—the most common one. Both it and orthoclase lack the fine lines or striations seen on the cleavage faces of plagioclase feldspars. Microcline is also found in a pale green color (amazonite, p. 88), sometimes of gem quality. The best crystals of amazonite are from granites near Pikes Peak.

ALBITE is one of the plagioclase feldspars. It is a sodium feldspar with a slightly lower specific gravity than others, and frequently contains potassium. The basal cleavage surface is marked with fine lines. Color white, gray, or bluish, often with a bluish sheen. Some are cut as moonstones. Albite is common in granitic rocks and in acidic lavas.

ANORTHITE is a calcium feldspar which ends the series beginning with albite. Between them is oligoclase, with 15 to 25 per cent calcium. Oligoclase is common in East Coast granites. Anorthite is less common; it occurs in several forms, mainly white, gray, or glassy. Small amounts of sodium are usually present.

crystal
Fort Bayard, N.Mex.

Pennsylvania

Albite twinning

Japan

Arkansas

Massachusetts

ACTINOLITE—green; fibrous or radiating, glassy crystals.

HORNBLENDE H. 5 to 6; Sp. Gr. 3 to 3.4. Crystals common.

Massachusetts

TREMOLITE—white, gray, or colorless; usually long, bladed crystals.

AMPHIBOLES are complex hydrous silicates containing calcium, magnesium, and iron. Crystals are often long or needle-like; sometimes fibrous (p. 75). The cleavage planes are at about 55° and 125°, forming wedge-shaped cleavage fragments. These characteristics are important in separating amphiboles from pyroxenes (p. 101). Hornblende, dark green to black and glassy, is found in basic igneous rocks and in such altered rocks as hornblende schist. It contains aluminum and is most often of secondary origin.

Actinolite and tremolite are similar minerals. The crystal forms are the same but in actinolite some of the magnesium is replaced by iron, giving the mineral a green color. Both are common minerals of sedimentary rocks related to amphibole jades and asbestos (p. 75).

Denmark

Arizona

crystal

crystals

ENSTATITE H. 5.5; Sp. Gr. 3.5. Color variable.

AUGITE H. 5 to 6; Sp. Gr. 3.5. Green to black; glassy.

Switzerland

DIOPSIDE H. 5 to 6; Sp. Gr. 3.4. Color, white to green and brown — sometimes transparent.

PYROXENES are complex silicates, closely related to the amphiboles. Pyroxenes are often found as primary minerals in igneous rocks. Their cleavage angles are close to 90°, giving squared cleavage fragments. They too are often fibrous or needle-like. Most are gray or green, grading into black. The kinds of pyroxenes are not easy to distinguish. They vary chemically as iron replaces calcium and magnesium. When this occurs, enstatite becomes hypersthene and diopside becomes hedenbergite. Enstatite is sometimes found in meteorites from outer space. Diopside, usually a light green, is most common in metamorphosed dolomitic marbles. Augite, the most common pyroxene, is a complex of aluminum, magnesium, calcium, and iron silicates found in nearly all basic igneous rocks and dark lavas, dikes, and sills.

ZEOLITES are not major rock formers but they are widely distributed. All are chemically related to the feldspars with the addition of water, chemically combined. This water is held loosely, so all zeolites boil and bubble when heated by a blowpipe. Their name means "boiling stone." About 60 minerals fit into the zeolite group. In addition there are several zeolite associates—minerals chemically similar but not of the zeolite structure. Zeolites and their associates are often found in lavas, filling cavities and veins. All are pale, fairly soft minerals of low density. The ability of zeolites to interchange ions of calcium and sodium has promoted the manufacture of artificial zeolites for use as water softeners.

Stilbite often occurs in pearly, sheaflike masses of twinned crystals. Radiating crystals, often translucent, may also form rounded knobs. Color: white, yellow, to reddish-brown. H. 3.5 to 4; Sp. Gr. 2.1.

Chabazite occurs with stilbite, usually in the form of large rhombohedral—almost cubic—crystals. It is white, occasionally pink, with a glassy luster; transparent or translucent. H. 4 to 5; Sp. Gr. 2.1.

Natrolite forms slender, prismatic needle-like crystals. It fuses in the heat of a candle flame—a distinguishing characteristic. Color, white to yellowish. H. 5; Sp. Gr. 2.2; luster, glassy.

Pectolite is seen in tapering masses of thin needles sometimes several inches long. H. 5; Sp. Gr. 2.8. Color usually white; luster silky. It is also fibrous or may form radiating masses. Pectolite needles are dangerously sharp and should be handled with care.

Prehnite and pectolite are zeolite associates which may occur together. Prehnite is usually in compact masses of flat light-green crystals. Luster, glassy; brittle and translucent. H. 6 to 6.5; Sp. Gr. 2.9.

STILBITE
New Jersey

CHABAZITE
Pennsylvania

NATROLITE—Oregon

PECTOLITE
New Jersey

PREHNITE—New Jersey

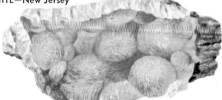

GARNETS are better known as gems than as rock-forming minerals, but they are common and form a small but conspicuous ingredient of igneous and metamorphic rocks. Garnets are a close-knit family of silicate minerals with many common characteristics. They all form crystals in the isometric system, usually with 12 or 24 sides, though sometimes combined forms with 36 or 48 faces are found.

Chemically, garnets contain the elements calcium, magnesium, iron, and aluminum, combined with silicon and oxygen. Other, less common metals may also occur.

in phyllite—Pennsylvania

GROSSULAR is a calcium-aluminum garnet, normally colorless to white, but colored when it contains iron as an impurity. It is found mainly in marble. A warm-brown variety from Sri Lanka is cut as a gem. The name refers to a Siberian "gooseberry green" variety.

cut gem   crystal

PYROPE is sometimes called precious garnet, though it is mined in large quantities for garnet paper. Gemmy, perfect specimens are found in South African "blue earth" with diamonds. Pyrope is a magnesium-aluminum garnet.

crystals in mica schist, New York

ALMANDINE is the "common garnet" found in many metamorphic rocks. It is an iron-aluminum garnet, though part of the iron may be replaced by magnesium—making such forms similar to pyrope. When it has a clear red color it is sometimes—like pyrope—called precious garnet and is cut as a gem.

Crystals are abundant—from pinhead size up to 4 in. in diameter. Fresh crystals have a glassy luster. There is no distinct cleavage. All garnets have a hardness of about 7. Their density is more variable, between 3.4 and 4.3—depending on the metals in them. Most garnets are found in schists, gneiss, and marbles. Some occur in lavas and in granites.

Only a small percentage of garnets are of gem quality. In a few large deposits, garnets are mined and crushed for garnet paper and other abrasives.

crystal in schist

Alaska

California

SPESSARTINE is quite a rare member of the group of aluminum garnets. It contains manganese and aluminum. The manganese often gives the garnet an orange tint which makes gem-quality specimens particularly valuable. Such gem-quality stones have come from Virginia, California, and Brazil.

ANDRADITE is a garnet containing calcium and iron. It, too, is very common and like almandine is called common garnet. Color varies from yellow to green, red, and black, depending on impurities. It occurs in igneous rocks and in some metamorphosed limestones. The green form, demantoid, is a gem (p. 85).

UVAROVITE is a less common garnet found in serpentine rock and in limestones associated with chromium ores. It is a calcium-chromium garnet. The chromium gives it its rich green color. It is unlike most garnets in that a splinter of it will not fuse when heated with a blowpipe.

crystal　　0.1 in.

Jackson County, N.C.

OLIVINE, also called chrysolite or peridot, is a magnesium-iron silicate, colored various shades of green (rarely, brown); H. 6.5 to 7; Sp. Gr. 3.3. Luster, glassy; transparent to translucent. Clear varieties are cut as the gem peridot. Olivine is found in igneous rocks that are rich in magnesium and low in quartz, as basalt and gabbro; also in metamorphosed dolomites. It is often found in the form of small grains or in large, granular masses. The crystals are relatively rare, though occasionally some have been found up to several inches long.

CHLORITE is the name given to a group of silicate materials that are composed of magnesium, aluminum, and iron with water. They often form as an alteration of rocks rich in pyroxenes, amphiboles, and biotite. It may also form in cavities of basic igneous rocks. Chlorite is usually green but may vary from white to brown and black. H. 2 to 2.5; Sp. Gr. 2.8; pearly luster; streak, greenish or white. It forms in masses, crusts, fibers, or bladed crystals. The crystals have a perfect basal cleavage and, like mica, split into thin sheets. These may bend slightly, like selenite, but are not elastic like mica.

Chester, Vt.

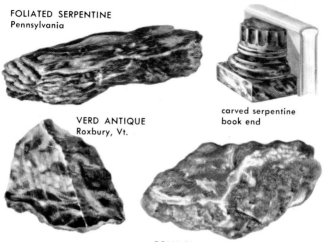

FOLIATED SERPENTINE
Pennsylvania

carved serpentine
book end

VERD ANTIQUE
Roxbury, Vt.

COMMON SERPENTINE—Norway

SERPENTINE (chemically similar to chlorite) is a group of magnesium silicates with water, but may include small amounts of iron or nickel. The fibrous form of serpentine, chrysotile asbestos, is described on p. 75. A number of other varieties depend on physical characteristics, especially color and luster. Common or massive serpentine $(H_4Mg_3Si_2O_9)$ varies from cream white through all shades of green to black. Streak is white. H. 2.5 to 4; Sp. Gr. 2.6; translucent to opaque. Note the greasy or waxy luster and feel of serpentine. Some weathered specimens are earthy. There are micaceous, fibrous, and mottled varieties.

The mineral serpentine is a secondary mineral which also occurs as metamorphosed serpentine rock. Deposits are large but at present have only minor use in firebricks. Serpentine marble (verd antique) and deeply colored common serpentine are used for carvings.

twinned crystals

New Mexico

STAUROLITE is an iron-aluminum silicate often found with garnets in such metamorphic rocks as schists, phyllites, and gneisses. Brown to black in color; streak, gray; H. 7 to 7.5; Sp. Gr. 3.7. Staurolite almost always occurs in crystals—as orthorhombic prisms and commonly as twinned crystals. Twinning may be at 60° or 90°. When at 90°, the twin crystals form a perfect cross. Such crystals (fairy crosses), broken or weathered from the bedrock, are sold as charms or souvenirs. These may be up to 2 in. long, but are usually an inch or less. Excellent specimens occur in schists near Mineral Bluff, Georgia. Transparent crystals occur rarely, and may be cut as gems.

EPIDOTE is one of a group of complex silicates of calcium and aluminum with water. It forms in nearly every type of metamorphic rock, in cracks and seams, as crystals or as thin green crusts. It is a typical mineral where igneous rocks have come in contact with limestones. Crystals are usually slender prisms, grading into needle-like forms. Color, green to brown and black; H. 6 to 7; Sp. Gr. 3.3. Easily identified by hardness and color.

crystal
Alaska

108

massive
California

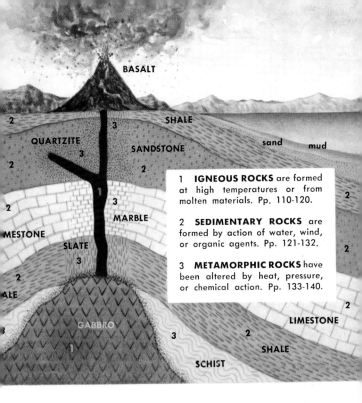

1 **IGNEOUS ROCKS** are formed at high temperatures or from molten materials. Pp. 110-120.

2 **SEDIMENTARY ROCKS** are formed by action of water, wind, or organic agents. Pp. 121-132.

3 **METAMORPHIC ROCKS** have been altered by heat, pressure, or chemical action. Pp. 133-140.

ROCKS

Rocks are large masses of material that make up the earth's crust. Some do not have discrete minerals but are composed of glasses or of organic materials like coal. A rock may consist of a single mineral such as quartz, gypsum, or dolomite. Most rocks contain several minerals, or were formed from older rocks in which these minerals were present.

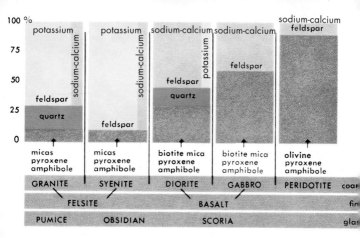

IGNEOUS ROCKS

Igneous rocks are classified by their texture, mineral content, and origin. They all come from magmas—molten mixtures of minerals, often rich in gases, found deep below the surface. If magmas cool beneath the surface they form *intrusive* rocks and develop typical structures that may later be exposed by erosion. Magmas reaching the surface form *extrusive* rocks, such as the spectacular volcanic rocks.

Igneous rocks usually contain ferro-magnesian minerals (amphiboles, pyroxenes, micas, or olivine) and feldspar or feldspar-like minerals. Many contain quartz. Those rich in light minerals (quartz and potassium feldspar) are called acidic. These are light not only in color but in weight (average Sp. Gr. 2.6 to 2.7). Those richer in ferro-magnesian minerals are called basic. They are darker and heavier (Sp. Gr. 3.0 and more). In texture, igneous rocks range from those with large crystals to glassy rocks with no crystals at all.

INTRUSIVE ROCKS form as magmas cool. This is a gradual process in which the more volatile chemicals remain as liquids and gases longer. Some intrusive rocks, very near the surface, grade into extrusive types. Those that cool deeper and slower are more coarsely crystalline.

Granite is the best known of the deeper igneous (plutonic) rocks. It is usually light-colored, formed mainly of potassium feldspar (about 60 per cent) and quartz (about 30 per cent), usually with mica or hornblende. The intergrown mineral crystals are all about the same size—a characteristic of slow, steady cooling. Fine granite has a salt-and-pepper pattern. Feldspar may redden it. Granite is hard and tough, widely used in construction and monuments.

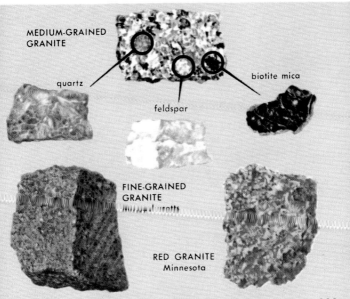

MEDIUM-GRAINED GRANITE

quartz

feldspar

biotite mica

FINE-GRAINED GRANITE

RED GRANITE
Minnesota

OTHER INTRUSIVE ROCKS include some which have cooled near the surface. These may contain a groundmass of crystalline grains surrounding larger crystals. Such igneous rocks made of crystals (phenocrysts) in a finer groundmass are known as porphyries. Porphyries are

Shelby, N.C.

PEGMATITE is a coarse-grained vein or dike rock with crystals that range from an inch or so to many feet in length. Pegmatites are mined for their mica and feldspar, or for gems and other accessory minerals. They often contain cavities or vugs lined with crystals. In one form, graphic granite, the quartz forms angular figures which look like writing.

Some pegmatite minerals:

garnet	arsenopyrite	spodumene
apatite	tourmaline	emerald
topaz	lepidolite	cryolite
beryl	chrysoberyl	sapphire
ruby	molybdenite	wolframite
pyrite	cassiterite	uraninite
fluorite		

Wausau, Wis.

SYENITE is less common than granite. It lacks quartz entirely or may have a small amount, in which case it is known as quartz-syenite. Syenite is mainly potassium feldspars with some mica or hornblende. The crystals are usually small and the rock is even-textured. Syenite also forms porphyries with phenocrysts of feldspar.

St. Cloud, Minn.

GRANITE PORPHYRY, which forms under somewhat different conditions from granite, has a granite groundmass in which phenocrysts of feldspar, quartz, or biotite mica are embedded. A porphyry is named after the matrix or groundmass—such as syenite porphyry, basalt porphyry. They also occur in extrusive rocks (p. 116).

found in intrusive rocks formed near the surface, and in extrusive rocks, but do not occur in deep-seated intrusives. Pegmatite, granite, and syenite are light-colored intrusives. Diorite, gabbro, and peridotite are dark, with more ferro-magnesian minerals.

DIORITE is a basic rock rich in minerals such as amphiboles, biotite, or pyroxenes. Its texture is like that of granite, but it is composed mainly of plagioclase feldspar and ferro-magnesian minerals. Other feldspars may be present, and sometimes a bit of quartz. Its color is usually gray or dull green. Granites grade into diorites through intermediate forms—the granodiorites.

Salem, Mass.

GABBRO, like diorite, has a granitic texture. Since texture depends upon rate of cooling, rocks of gabbro composition vary from fine-grained (diabase) on the outside of a dike or sill (p. 114) to a typical gabbro within. Gabbro is made mainly of plagioclase feldspar and pyroxene, with some olivine, traces of ilmenite, but no quartz. A dark rock, its mineral crystals are deeply intermeshed, making it a very tough rock. Porphyritic gabbros are rare.

Dolerite (or diabase) is the fine-grained gabbro which often occurs in sills or dikes. It grades into basalt at the edges, where it has cooled more rapidly.

Salem, Mass.

PERIDOTITE is a dark, heavy intrusive rock composed mainly of olivine with pyroxene and tiny flecks of phlogopite mica or hornblende. Little or no feldspar is present. Fresh rock is nearly black, but the more common weathered specimens are greenish and softer. Peridotite alters into serpentine. The South African diamond deposits occur in peridotite, which, in other places, contains important deposits of nickel, chromium, and platinum.

Baste,
Harz Mts., Germany

INTRUSIVE ROCK STRUCTURES are the natural forms taken by intrusive rocks. Sometimes these structures form deep beneath the surface. With the passing of time, they may later be exposed as covering rocks are removed by water, ice, or wind. When intrusive structures appear at the surface they may become spectacular features of the landscape. As they weather they produce typical kinds of soil. Rich ore deposits formed with their intrusion may then become available.

Palisades of the Hudson
(scale distorted)

DIKES are sheetlike intrusions rising from a batholith (p. 115) or from some other source of magma. These sheets may vary in thickness from a few inches to hundreds of feet, and may extend from a few feet to many miles. The magma forming the dike follows cracks and joints below the surface, hence dikes characteristically cut across the rock structures. Magma cools rapidly in contact with the surrounding rock. Hence the dike may differ in texture and composition in these contact zones. The heat of the injected material may metamorphose the adjoining rock (p. 139). Miles of well-developed dikes are exposed near Spanish Peaks, Colorado.

SILLS are similar in origin to dikes. While dikes cut across the existing rocks, sills form parallel to them, frequently shouldering their way in between layers of sedimentary rocks. If these layers are tilted, the sill will be tilted also. The Palisades along the Hudson River near New York is a tilted sill of diabase and gabbro. Sills and dikes may occur near volcanoes when lava fills up cracks in previous lava flows. Joints forming regular 5- or 6-sided columns mark some fine-grained intrusive rocks, as in the Giant's Causeway in Ireland and Devil's Tower in Wyoming.

Intrusive structures originate from deep reservoirs of magma whose origins are still not clear. Magmas are capable of producing both acidic and basic rocks. Acidic rocks tend to form major structures—batholiths, the cores of mountain ranges, and the great continental shields. Basic rocks occur more frequently in dikes and sills, often grading into extrusive rocks. All extrusive rocks, at one place or another, grade or join into intrusives. Recognizing such structures enables one to interpret the landscape.

LACCOLITHS might be likened to blisters within the earth's thin skin of sedimentary rocks. Magma spreads outward between rock layers and raises those above it into a dome that may be a thousand feet or more high and from one to ten miles across. Sedimentary layers lifted by the magma may crack, subjecting them to a more rapid erosion which may eventually expose the igneous core. The edges of resistant upturned sedimentary layers around the laccolith may form circular ridges called hogbacks. Laccoliths are common in the West, particularly in Utah.

BATHOLITHS, the largest intrusions, may cover 100,000 square miles. Our largest, in central Idaho, spreads over 16,000 square miles. In the Rockies and Sierra Nevada batholiths are exposed in the mountain cores. Though some are ancient, some are new, indicating that they are part of the continual building up of the earth's crust. As magma wells up it may dissolve some of the surrounding rocks. It may spread unevenly, trapping islands of older rock. Batholiths generally contain coarse-grained rocks. They are deeply buried and may not be uncovered for millions of years. Small batholiths are called stocks.

EXTRUSIVE ROCKS embrace a large group, the most common member of which is lava. This molten material pours out through fissures and volcanoes. It spreads in great lava flows or builds up cones. Volcanic explosions throw fragments into the air as volcanic ash or volcanic bombs. The forms and structures of these rocks are shown

RHYOLITE—Colorado

RHYOLITE is a light-colored acidic rock which has very much the same chemical composition as granite. Its texture is very fine. When minerals can be seen in it—phenocrysts of quartz and a glassy feldspar (sanidine)—the rock is a rhyolite porphyry. Color white to pink to gray, though often reddish from iron stains.

OBSIDIAN—Wyoming

OBSIDIAN and PUMICE are chemically the same as rhyolite. Obsidian or natural glass is formed when rhyolitic lava is quickly chilled. Though it is dark-colored, thin fragments are light and transparent. Indians made knives, arrowheads, and ornaments from this unusual rock. Pitchstone is a duller, rougher form of obsidian. Rhyolite lava blown to a spongelike consistency by the release of gases forms pumice—a volcanic froth. Pumice is so light it floats on water, and fragments may be washed ashore far from the volcano.

PUMICE—Millard Co., Utah

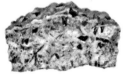

ANDESITE

ANDESITE is named for the Andes Mountains, where it is abundant. It contains little or no quartz, and has a greater proportion of ferro-magnesian minerals, which give the rock a darker color than rhyolite. Porphyries are common, with phenocrysts of feldspar or dark minerals. Andesite is intermediate in composition between rhyolite and basalt. Since rhyolite and andesite are difficult to distinguish in the field, the term felsite is used for both when more accurate identification is impossible.

on pp. 118–119. In color the rocks themselves may be light (acidic) or dark (basic). Like intrusive rocks, the extrusives grade from those rich in quartz to those with no quartz at all. Most of the rocks are fine-grained because of rapid cooling, and are difficult for the amateur to identify and classify.

BASALT is the common dark, dense lava that is widespread the world over. It is mainly pyroxene and a plagioclase feldspar, but the texture is so fine that these individual minerals are rarely seen. Olivine may also be present. In total, the rock is about half feldspar, half ferro-magnesian minerals. Basalt varies from a dark gray with a greenish tinge to almost black. In arid regions exposed basalt surfaces frequently develop a white, limy encrustation. In humid areas the iron in basalt oxidizes, coloring the surface a rusty brown.

Somerset County, N.J.

In addition to the dense rock found in lava flows, dikes, and sills, basalt has several other forms. Upper surfaces of basaltic flows may contain gas bubbles which form a porous and cindery rock called scoria. Later the holes or vesicles may become filled with minerals such as calcite, agate, and amethyst, as well as zeolites. When the openings are almond-shaped the name amygdaloidal basalt is used. Large deposits of native copper have been found in such basalts, especially in Michigan.

arid weathering

humid weathering

SCORIA
Klamath Falls, Ore.

AMYGDALOIDAL BASALT
Michigan

117

1 cinder cone
2 spatter cone
3 shield lava cone
4 compound volcano
5 radiating dike
6 lava flow

VOLCANOES AND LAVA FLOWS are the structures most typical of extrusive rocks. The Columbia lava plateau, consisting of a large number of lava flats, covers over 200,000 square miles; in places it is up to a mile thick. Volcanic cones vary from steep cinder cones (1) rarely 1000 ft. high to smaller spatter cones (2) and low, broad shield lava cones (3) with slopes of only a few degrees. Most volcanoes are compound (4), with cinders, ash, and broken fragments (breccias) interbedded with lava (see below). Some have radiating dikes of basalt (5); others are associated with extensive lava flows (6).

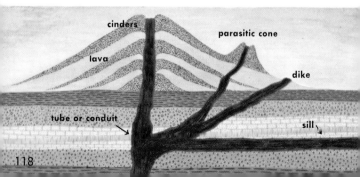

Pahoehoe is the Hawaiian name for a fluid lava which flows freely and cools with a smooth, ropy surface. Such lava flows look like frozen rivers of black, tinted with iridescent purple.

Hawaii

Viscous Lavas form crusts which break and roll over. This jumble, mixed with scoria, forms sharp, jagged flows. Such lava is called aa, another Hawaiian name. Volcanic lavas have been studied intensively at Mauna Loa and Kilauea, both shield cones.

Hawaii

Volcanic Bombs are masses of liquid lava thrown into the air. Their motion gives them the elongate shape and smooth surface. Active volcanoes throw vast quantities of gases and steam into the air as well as dust, ashes, and fragments ranging from lapilli (less than 1 in.) to blocks weighing tons.

Craters of the Moon, Idaho

Lava Caves and Tunnels form when a strong crust hardens under which the liquid lava keeps flowing, leaving a hollow. Crude caves form in overturned masses of aa lava. In some, dripping lava forms stalactites. Ice and snow which fill some lava caves in winter are so protected that they do not melt by summer — making "ice caves," a tourist attraction.

New Mexico

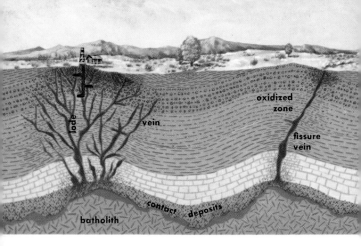

IGNEOUS ROCKS have long been known to be associated with metal ores. Hot liquids and gases from magmas cool to produce ores directly, or form ores as they react with the local rocks they penetrate. Contact deposits are formed around the edges of batholiths or other large intrusions, especially when these have penetrated into limestones. Veins may extend from igneous masses into the local rock, carrying mineralizing liquids and gases. A system of veins which can be mined as a unit forms a *lode,* like the famous Mother Lode of California gold days. Other veins form as mineralizing solutions penetrate cracks and deposit ores in natural openings or zones of crushed rock.

Most primary ores are sulfides or oxides. These and other ores often occur in specific mineral associations. One example is the lead-zinc deposits of Kansas and Oklahoma (sphalerite, galena, and many accessory minerals), another is the zinc, iron, and manganese deposits at Franklin, N.J.; a third is the iron-titanium ores of the Adirondacks.

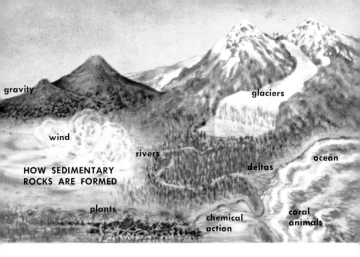

Diagram labels: gravity, glaciers, wind, rivers, ocean, deltas, **HOW SEDIMENTARY ROCKS ARE FORMED**, plants, chemical action, coral animals

SEDIMENTARY ROCKS

SEDIMENTARY ROCKS are extremely varied, differing widely in texture, color, and composition. Nearly all are made of materials that have been moved from a place of origin to a new place of deposition. The distance moved may be a few feet or thousands of miles. Running water, wind, waves, currents, ice, and gravity move materials on the surface of the earth by action that takes place only on or very near the surface. In total these rocks cover about three-quarters of the earth's surface.

Unconsolidated mud or sand is usually referred to as a sediment, while consolidated materials are called sedimentary rocks. Rocks made up of grains or particles are called clastic; they may range from less than a thousandth of an inch to huge boulders. Other sedimentary rocks are of chemical or organic origin. Most sedimentary rocks form in layers or strata; many contain fossils. Major sedimentary strata form slowly over millions of years.

Great White Throne (sandstone) at Zion National Park, Utah.

SANDSTONE is formed by the action of wind, water, and ice on older rock. It is mainly grains of quartz cemented by silica, lime, or iron oxide. Silica cement may produce hard, durable sandstones; the other cements are not as resistant. Sandstones grade off on the coarse side into conglomerates (p. 128) and on the finer side into sandy shales (pp. 124–125). Most sandstones formed in shallow seas and show signs of near-shore origin—they often include fossil ripplemarks and shells of shallow-water animals. In some sandstones, nodules of soft goethite or hematite form "paintpots" once used by American Indians.

MEDIUM-GRAINED SANDSTONE
New York

SANDSTONE
cemented by iron oxide—Utah

ARKOSE—showing feldspar grains
Mt. Tom, Mass.

Ripple marks in sandstone
Ohio

"Indian paint pot"
Glen Cove, N.Y.

Cross-bedded sandstone formed from ancient dunes—Kanab, Utah.

CALCAREOUS SHALE
Livingston County, N.Y.

SANDY SHALE
Kent, N.Y.

SHALES or mudstones are mainly clays which have hardened into rock. Shales may grade into fine sandstones or, when much lime is present, into shaly limestones. Since clay particles are exceedingly fine, they tend to be carried into deep or quiet water. Shales are often thin-bedded or laminated, with fairly uniform texture. Their color is usually gray, but varies from black to dull red.

Clays and shales are frequently so fine-grained that they serve as a barrier to movements of water. Clays and silts that form in lakes may show dark and light alternating layers called varves, each pair of layers representing a year's deposit. By counting the varves the age of glacial lakes and other deposits can be estimated. For instance, varve counts indicate that the shales of the Green River formation in Wyoming took over 5 million years to form. Other structures found in shales are shown on the next page. Shales are used in the manufacture of cement. Unconsolidated clays (p. 150) are of tremendous value for ceramic and other uses. Since clays and shales are commonly formed from rocks rich in feldspar, they contain much aluminum silicate. Some shales are oil reservoirs. Billions of tons of oil shale are a potential source of petroleum for future use—when present high-yield sources will have been depleted.

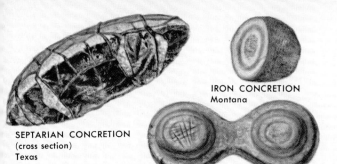

SEPTARIAN CONCRETION
(cross section)
Texas

IRON CONCRETION
Montana

CONCRETION—Sandusky, Ohio

CONCRETIONS are often found in shales, sandstones, and limestones. They may be spherical or flattened masses formed around a fossil or some other nucleus. They may be very small or up to several feet in diameter. Concretions are often harder than the enclosing rock and hence are found as they weather out of it.

MUD CRACKS form as deposits of mud and clay in shallow lakes or on mudflats dry in the sun. The shrinkage forms rough six-sided blocks. Later fresh mud may be washed into the shrinkage cracks. Mud cracks indicate the shallow-water origin of the rock.

consolidated
mud cracks

raindrop imprints in shale
Illinois

RAINDROPS falling on mud or clay at just the right consistency will leave small pits. Continued rain would wash these away, but a brief shower followed by a period of dryness may preserve raindrop impressions. These are rarer and harder to find than mud cracks.

LIMESTONES are exceedingly variable in color, texture, and origin. They consist mainly of the mineral calcite (p. 63) and react like calcite chemically. Most of them are of marine origin—some forming at great depths. Both plant and animal life, directly and indirectly, contribute to their formation. Many kinds of animals contribute the minerals that form limestone—corals, worms, crinoids, mollusks, and certain protozoa. Algae are also important lime-precipitating plants. Some limestones are chemical precipitates, while others are cemented fragments of lime.

A unique and complex natural balance is known to involve the carbon-dioxide content of the air, the carbon-dioxide and lime dissolved in the sea, limestone formation, and climatic change. Limestone rocks are a keystone in this structure because they are great reservoirs of carbon dioxide as well as of lime. Since lime is soluble in acid water, limestones dissolve and recrystallize easily. They

Florida

SHELL LIMESTONES include *coquina*, formed recently of shells and fragments loosely cemented. *Chalk* is a limestone made of tiny protozoan shells. Older shell limestones may contain fossil brachiopods, bryozoans, corals, and crinoids. Some of the oldest known sedimentary rocks are limestones formed from algae.

Indiana

OOLITIC LIMESTONES usually consist of a mass of small concretions, each built up layer upon layer around some small nucleus. The resulting rock is composed of spherical grains. Oolitic limestones may be formed in shallow water. Each grain grows as it is rolled by waves or currents.

weather rapidly in humid climates; very slowly in arid ones, where they are good cliff formers—as at Grand Canyon.

Limestones vary from almost unconsolidated masses of shells (such as the kind shown on p. 126) to compact, crystalline rocks. Intimately connected with plant and animal life, they are a rich source of fossils (pp. 130–132). Limestones rich in clay are known as *marls*. They also grade into shales and into sandstones. Some contain silica concretions in the form of chert, flint, or chalcedony (pp. 80–81). A carbonate rock which contains calcium and magnesium carbonate is a dolomitic limestone or a true dolomite (p. 65).

The economic importance of limestones is almost beyond estimation. Limestones are widely used for road metal, in concrete, and for lime. Shaly-limestones are a source of cement. Lead, zinc, fluorite, sulfur, and oil deposits are often associated with limestones.

TUFA is a light, porous limestone often colored with iron. It forms in springs, where calcite may be deposited on water plants, twigs, or debris. In caves, the secondary deposit of calcium carbonate (*travertine*) forms flowstone, covering walls and floor and sometimes forming stalactites and stalagmites (pp. 63–64).

CRYSTALLINE LIMESTONES form as calcite recrystallizes to a greater or lesser extent. These limestones can closely alter in marble (p. 134) but are considered sedimentary rocks when there is no indication that they have been deformed by pressure. Crystalline limestones used for decoration and ornamentation are sometimes sold as marble.

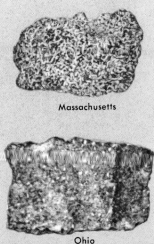

Massachusetts

Ohio

CONGLOMERATE
New York

CONGLOMERATES are sedimentary rocks composed of rounded pebbles one-fourth of an inch in diameter or larger, cemented in a matrix of finer material. In tillite, a conglomerate of consolidated glacial till, the unassorted fragments may range from gravel size to great boulders. The larger ones may show typical glacial scratches. The coarse materials which form conglomerates are deposited close to shore at the mouths of swift rivers or canyons, in alluvial fans, or in deltas. The pebbles in conglomerates are frequently quartz or quartzite. The cement may be iron oxide, silica, calcium carbonate or, occasionally, clay. Conglomerates grade off into sandstones.

Conglomerates in which the fragments are sharp and angular because they are freshly broken and not worn in transport are called breccias. The various kinds of breccias have little in common other than the angular shape of their components. Some are of volcanic origin, some represent cemented materials in talus slopes or alluvial fans, and some have formed along fault zones. Breccias may be barely consolidated or tightly cemented, depending on their age and on conditions of formation. Commonly they are formed close to the point of origin of the fragments.

QUARTZ BRECCIA
from Texas—fragments cemented by silica.

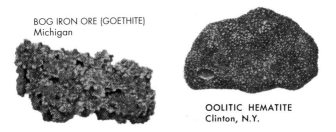

BOG IRON ORE (GOETHITE)
Michigan

OOLITIC HEMATITE
Clinton, N.Y.

CHEMICALLY FORMED sedimentary rocks are perhaps the most important commercially. Some limestones are precipitated chemically from sea water. Anhydrite, gypsum, and halite (pp. 66–69) form deposits large enough to be considered chemical sedimentary rocks as well as minerals. They are often residues from sea water, which contains about 3½ lbs. of chemical solids in every 100 lbs. of water. Such deposits generally indicate an arid climate at the time of their formation.

Other sedimentary rocks and fuels are formed through biochemical action, and of these coal and oil (pp. 142–145) are the best known. Bacteria may aid in the formation of iron deposits in swamps and shallow lakes, and may be in part responsible for the great Iron Range deposits. Biochemical action may account for the precipitation of manganese also. Diatomaceous earth is a fine deposit of microscopic plant skeletons—forming large, pure deposits of silica.

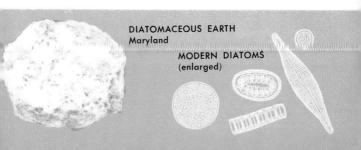

DIATOMACEOUS EARTH
Maryland

MODERN DIATOMS
(enlarged)

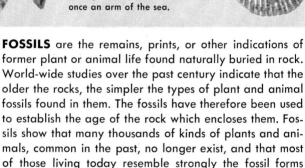

Arrowheads found buried offer evidence of human life. At left, an ancient folsom point from New Mexico. At right, marine fossils found along Lake Champlain, N. Y., show that this lake was once an arm of the sea.

FOSSILS are the remains, prints, or other indications of former plant or animal life found naturally buried in rock. World-wide studies over the past century indicate that the older the rocks, the simpler the types of plant and animal fossils found in them. The fossils have therefore been used to establish the age of the rock which encloses them. Fossils show that many thousands of kinds of plants and animals, common in the past, no longer exist, and that most of those living today resemble strongly the fossil forms found in relatively recent rocks.

In addition to telling the details of life in the past and the story of such unique animals as giant dinosaurs and titanotheres, fossils also tell of past climates. Colonial corals in Greenland rocks attest to warmer conditions in the past than today, and imprints of fir and spruce in unconsolidated clays near the surface record the penetration of glacial cold far to the south. Fossils are also used to determine the marine or fresh-water origin of rocks.

The occurrence of fossils is both rare and common. Only a tiny fraction of the total number of living things has ever been preserved as fossils, and yet certain layers of rock or strata are made almost entirely of shells, teeth, plant remains, and even of bone.

FOSSILS are preserved in many ways. The simplest is the intact preservation of the hard parts of a plant or an animal, as illustrated on p. 130. Wood, bone, teeth, and other hard parts are preserved intact for relatively short periods.

In another type of fossilization, buried plant or animal materials decompose, leaving a residual film of carbon behind. This may mark the form of a leaf or of some simple animal. On a larger scale this process is responsible for our great deposits of coal.

Sometimes buried material is gradually replaced by silica and other material like calcite, dolomite, or pyrite from solutions which permeate the rock in a process called petrifaction. These replacements form another very common type of fossil.

Probably the most spectacular of all replacements is that of wood by agate or opal as a result of the action of hot, silica-bearing waters. This forms petrified wood. The replacement may be so minute and complete that even the details of cellular structure are preserved. The best-known examples are preserved in the Petrified Forest National Park in Arizona.

CARBONIZED FERN LEAF
Illinois

PETRIFIED WOOD
Arizona

cell
structure
(enlarged)

CAST OF SHELL
in pyrite
western Illinois

131

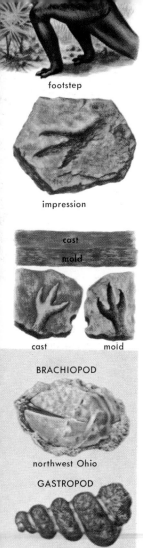

footstep

impression

cast

mold

cast   mold

BRACHIOPOD

northwest Ohio

GASTROPOD

northern Ohio

MOLDS AND CASTS are very common fossil forms. They are impressions, and so differ from intact preservation and replacements. A footprint, as that of a dinosaur, is a good example of a mold. The impression left in soft mud or silt may harden before more sediment fills it in and provides material for a new layer of rock. If the sediment later consolidates and the rock is eventually broken open, the original imprint will be found below, and filling it will be a cast of the underside of the dinosaur's foot.

When shells are buried in sand or mud, a mold of the outer surface of the shell is formed. Percolating waters may dissolve the shell material completely, and the mold will then be the fossil record. Later, percolating waters may refill the cavity with calcium carbonate or silica, forming a cast which will on its outer surface completely duplicate the external form of the shell.

Paleontology, the study of plant and animal fossils and their histories, is an important and exciting branch of the science of geology. The study of small fossil forms (which are known as microfossils) has recently yielded a great deal of new information.

METAMORPHIC ROCKS

Metamorphic rocks are rocks which have been changed. Changes may be barely visible, or may be so great that it is impossible to determine what the original rock once was. All kinds of rocks can be metamorphosed—sedimentary, igneous, and other metamorphic rocks. The changes usually bring about a new crystalline structure, the formation of new minerals, and sometimes a coarsening of texture.

Metamorphism results from heat, pressure, or permeation by other substances. Pressure and heat increase with depth in the earth's crust (A, in the illustration above) and may also result from crustal movements (B) or igneous activity (C). Rocks may be permeated by gases or fluids from igneous material (D) or by the percolating of mineral-bearing ground-water.

SIMPLE METAMORPHIC ROCKS is a convenient term applied to rocks formed by the direct alteration of sedimentary rocks where the changes are mainly recrystallization. Few, if any, new minerals are formed. Some of these rocks show parallel structures; others do not.

Slate results from metamorphism of shale, and often traces of the original bedding can be seen. Slate is frequently of a blue-gray color but may be green, red, or brown. It breaks easily along a flat cleavage plane and can be split into sheets used for roofing or flagstones. Sometimes slate shows folding and wrinkling.

Marbles are recrystallized limestones, normally white, but often tinted by iron oxide, carbon, or serpentine to attractive shades of yellow, brown, green, or black. Limestones and dolomitic limestones may be slightly altered by percolating waters and are often called marbles, but true marbles are the result of metamorphism involving heat and pressure. Secondary minerals may form and crystals may show distortion. Marbles do not often develop the parallel banding and mineral arrangement seen in slates and schists.

Quartzites are usually metamorphosed sandstones which have recrystallized so that, in breaking, they break through the quartz grains instead of through the cement, as in sandstones. The grain structure in quartzite is not nearly as clear as in sandstones. Quartzite, like marble, is a massive metamorphic rock, very hard and tough.

Hornfels are clays or shales which have been metamorphosed through the action of heat from nearby igneous rocks. The hard, recrystallized rock may retain its sedimentary structure, but garnet and other secondary minerals may form. The color is usually dark; the rock is spotted or banded, and may be confused with basalt, especially along contacts.

GRAY SLATE
Pennsylvania

RED SLATE
New York

WHITE MARBLE
Georgia

BLACK MARBLE
Italy

HORNFELS
Hartmannsdorf, Saxony

QUARTZITE
Dell Rapids, S.Dak.

PHYLLITES AND SCHISTS represent the more highly metamorphosed rocks. Phyllite provides the transition, being more metamorphosed than slate but less than schists. Fine grains of mica give it a silky luster. The schists are coarser than phyllite, with a considerable amount of mica or other secondary minerals. They break in a wavy, uneven surface; this property is called *schistosity*. Schists are named for their most characteristic mineral:

Mica Schist is usually a highly metamorphosed shale composed mainly of many small flakes of mica, oriented roughly parallel, and quartz. Texture varies from fine to coarse, and either staurolite or garnet may be present.

Hornblende Schist is mainly hornblende and quartz. It is dark in color, and while the minerals have a parallel orientation, hornblende schist does not break cleanly.

Chlorite Schist contains chlorite as a metamorphic mineral instead of mica; this gives it a greenish color. It has typical schist orientation.

Quartz Schist forms on further metamorphism of impure quartzite. Muscovite mica usually develops as a secondary mineral, and parallel structures also form. Light-colored.

MINERALS OF METAMORPHIC ROCKS

The following minerals are commonly found in metamorphic rocks.

Actinolite	in schists, gneiss and quartzite (p. 100)
Chlorite	in phyllites and schists (p. 106)
Diopside	in marbles (p. 101)
Feldspars	mainly from igneous contacts (p. 98)
Garnets	nearly all kinds— in schists, marble, and phyllites (p. 104)
Graphite	in some schists and marble (p. 62)
Hornblende	in metamorphosed basic rocks (p. 100)
Kyanite	in schists (p. 48)
Micas	muscovite, biotite, and phlogopite in schists, gneiss, and marbles (p. 96)
Olivine	in marbles (p. 106)
Quartz	in schists, gneiss, and quartzite (p. 76)
Serpentine and talc	(p. 74) in marbles, soapstones, and schists (pp. 74 and 107)
Staurolite	in schists (p. 108)

Page numbers indicate illustrations.

PHYLLITE
Connecticut

GARNET-MICA SCHIST
New York

HORNBLENDE SCHIST
Lingoutte, Vosges

CHLORITE SCHIST
Chester, Vermont

QUARTZ SCHIST
Pennsylvania

137

MUSCOVITE GNEISS
Nassengrub, Bohemia

GRANITE GNEISS
Bristenstock, Switzerland

GNEISS (pronounced nice) may be simply metamorphosed granite, or a far more complex rock with possibly four or five different origins, either igneous or sedimentary. It may also include metamorphic rocks which are invaded by igneous materials so that the rock becomes a complex mixture (migmatite). Schists are often invaded in this way, producing rocks which contain more feldspar and quartz than ordinary schists. The new minerals are often in small lenslike intrusions. Gneiss is hard to define or describe because it is so varied. In general, it is a coarse-textured rock with the minerals in parallel streaks or bands, but lacking schistosity. It is relatively rich in feldspar and usually contains mica or one of the other dark, rock-forming minerals.

Gneiss is classified by its most conspicuous mineral or according to its origin or structures. Characteristics are usually better seen in the field than in hand specimens.

HORNBLENDE GNEISS
Stengerts, Bavaria

INJECTION GNEISS
Sewen, Vosges Mts., France

Muscovite Gneiss is one of the most common kinds, with a pale salt-and-pepper appearance, though biotite mica is common in gneiss also. While the name gives no clue as to the origin, muscovite and biotite gneisses may form from highly metamorphosed, shaly sediments.

Granite Gneiss is named to indicate that it is a metamorphosed granite, at one time thought to be a product of granitization (p. 140). Granite gneiss is rich in feldspars. The mica or hornblende in it is arranged in parallel bands.

Hornblende Gneiss is a dark rock, much darker than biotite gneiss, in which parallel-oriented hornblende replaces mica. It is probably the end result in the metamorphosis of basic igneous rocks.

Injection Gneiss is gneiss which has been permeated by igneous materials during metamorphism. Like schists which are thus altered, it is also known as migmatite.

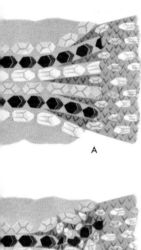

A

B

C

GRANITIZATION was once thought to be responsible for the formation of granite. It was suggested that some form of igneous material invaded sedimentary or metamorphic rock, creating mixtures which eventually altered the rock so that it becomes like granite in texture and composition. This process was associated with great batholiths (p. 115) or materials coming from unknown depths beneath the crust. Gradations of rock from granite to gneiss were thought to be the primary evidence for the process of granitization. Today these gradations are believed to be caused by the assimilation of the gneiss by the granite due to the melting that results from the positioning of the batholith.

One example of how this process was alleged to have occurred would be the invasion of gneiss (A) by solutions containing quartz and feldspar that divide gneiss along parallel bands. In a later stage (B) some parts of the gneiss are transformed into granite while others retain their original structure. As granitization continued, the form and structure of the gneiss minerals change (C), though traces of the parallel bands still remain. Finally the original rock is completely absorbed. No traces of gneiss remain. In structure and composition it has become granite.

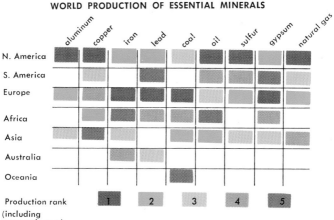

WORLD PRODUCTION OF ESSENTIAL MINERALS

ROCKS AND MINERALS IN DAILY USE

Rocks and minerals are the basis of our civilization. For the life we live today we must have metallic and non-metallic ores, the fuels, and the construction materials such as cements, clays, sand, gravel, and building stones. Then there is the soil, perhaps the most valuable material of all. Sand, clays, road rock, and soil have a low value per ton compared to metallic ores, but because of the amounts used their overall value is enormous. Since ores have been discussed earlier, this section deals mainly with fuels, soils, and construction materials.

Much of the trouble between nations can be traced in the fact that the rocks, minerals, and fuels are not equally distributed over the earth. Some nations have; others have not. Blood has been shed over gold, silver, iron, coal, oil, and uranium. This is still a critical matter for us all.

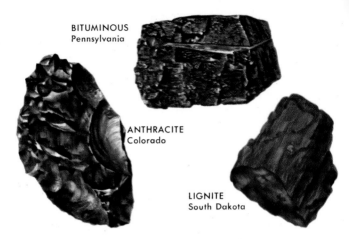

BITUMINOUS
Pennsylvania

ANTHRACITE
Colorado

LIGNITE
South Dakota

COAL is the fuel that made the industrial revolution possible, and it still is widely used. About a million tons are mined annually in the United States. Coal is an organic sedimentary rock consisting of the altered remains of plants. It is formed by a slow series of changes marked by a loss of water and volatile substances and a corresponding increase in the amount of "fixed carbon." Coal is classified by the relative amount of these three groups of materials. Peat, which contains about 80 per cent moisture, is not considered a form of coal.

Lignite, the lowest rank or kind of coal, has a heating value of 7,400 British Thermal Units (B.T.U.). It is brown in color and breaks down into powdered or flaky fragments when stored. It burns with a smoky flame.

Bituminous or soft coals are black coals which often have a cubic fracture and a dull luster. They yield from 9,700 to 15,400 B.T.U. and total 90 per cent of the coal mined. The higher grades of bituminous coal store well and burn with an almost smokeless flame.

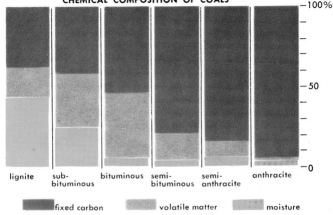

CHEMICAL COMPOSITION OF COALS

lignite sub-bituminous bituminous semi-bituminous semi-anthracite anthracite

fixed carbon   volatile matter   moisture

Anthracite or hard coal is hard and durable. It stores well and burns with a very short smokeless flame. Anthracite has a conchoidal fracture and a black shiny luster—sometimes iridescent. It forms when folding or metamorphism drives a larger amount of volatile matter out of soft coal than would otherwise be lost.

Most coals in eastern and central United States were formed during the Pennsylvanian period when a warm climate favored the rapid growth of fernlike plants. The coals of western United States were formed many millions of years later. The coal-bearing rocks are sometimes thousands of feet thick, with layers of coal up to 100 feet thick lying between layers of sandstones and shales.

The world's reserves of coal are very roughly estimated at a total of about seven trillion tons, of which over half is in North America. Not all of these reserves are usable at the present price of coal.

143

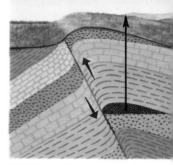

ANTICLINES or upward folds provide traps for oil and gas. Occasionally the surface pattern indicates underlying structures.

FAULTS mark movement along breaks in the earth's crust. Oil may seep to the surface or accumulate along or against them.

PETROLEUM is the proper name for what most people call oil. It is a complex mixture of compounds of carbon and hydrogen, for which new uses are constantly being found. The commercial production of petroleum began in the 19th century. Now the United States produces over two billion barrels a year and imports oil in addition to this. Nearly all internal combustion engines depend on petroleum fuels and lubricants.

Petroleum is of marine origin and probably represents the remains of microscopic plants which settled in the sand and mud of shallow bays. Deposits of oil bearing sands and shales are plentiful, but folding, faulting, or other action in the sedimentary strata (as illustrated above) was necessary to form the structures for the accumulation of the petroleum. Such accumulations are not underground lakes of oil, but are areas where the spaces between sand grains or the pores in carbonate rocks are saturated with petroleum.

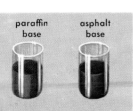

paraffin base   asphalt base

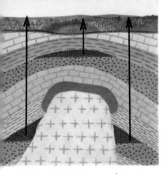

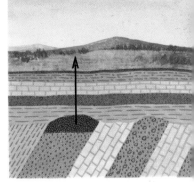

SALT DOMES, pushing up through Gulf Coast sediments and deforming them, make suitable structures to trap oil.

UNCONFORMITIES Oil may be confined where tilted layers have been worn down and later horizontal layers deposited on them.

The most valued oils have a paraffin base—that is, the heavier "chain" type chemicals are those of paraffin. Another form of petroleum has an asphalt base and is richer in "ring" type chemicals. Many wells yield a mixture of both these types, and chemically the oils contain many complex hydrocarbon compounds. On one hand, petroleum compounds grade off into natural gas (p. 147). On the other, they grade into the solid bitumens and natural asphalt.

Because of its great value, petroleum has stimulated more geologic exploration and research than any other mineral or rock. Petroleum surveys extend into wilderness areas and even out to sea. Tremendous reserves which can be recovered economically have been discovered. As a by-product of petroleum research our knowledge of sedimentary rocks and of their structures has been greatly increased.

Oil shale with
fossil fish
Wyoming

Oil shale
Menard County, Ill.

OIL SHALE contains solid hydrocarbons mixed with plant remains. Extensive deposits of gray, brown, or black oil shale are found in the Green River formation of Colorado and in Utah, Wyoming, and Western Canada. Oil shale has been mined and used for some time in Scotland. In the United States it cannot compete with our rich oil "pools." Should our richer deposits become depleted, we have well over 100 billion barrels of oil locked up in oil shales. During World War II an experimental plant successfully made oil from oil shale. A ton of average oil shale, upon heating, yields about 25 gallons of petroleum, nearly 10,000 cubic feet of natural gas and ammonium compounds.

It is certain that folding and other crustal movements produced most of the traps in which oil accumulated. The oil and gas apparently migrated from a source rock through the porous sands until it was trapped against an impervious surface and slowly accumulated. Without these crustal traps petroleum may remain distributed in shales and sands, too diffuse to be produced economically at

Courtesy of Mines Magazine

present prices. Oil shales and also oil sands contain huge reserves of solid hydrocarbons, from which oil can be made. How they will be utilized is, at the present moment, still an open question.

NATURAL GAS is found with petroleum, though some oil fields have very little gas and some gas fields yield no commercial oil. Chemically, natural gas is a mixture of the lighter chemicals found in petroleum—mainly methane with butane, propane, and other gases. Carbon dioxide, nitrogen, hydrogen sulfide and even helium may be present also.

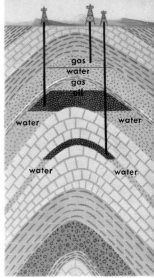

Production of natural gas in the United States averages about 540 trillion cubic meters annually. Reserves are enormous. Over 75 per cent of natural gas goes into industrial uses. Some is liquefied as bottled gas, and natural gasoline is an important by-product. From 1½ to 2 gallons of gasoline may be obtained from 1000 cubic feet of nat-

WELL FIELD producing gas from the upper zone, oil and gas from the middle zone, and oil below.

ural gas. Natural gas gasoline is produced at the rate of over 5 billion gallons a year, accounting for nearly half of all U.S. gasoline. Compression plants at the wellheads extract the gasoline from the "wet" gas. Over a quarter of a million miles of pipelines, some more than a thousand miles long, carry natural gas and petroleum from wells to industrial centers and cities.

In former times natural gas pressure on a pool of oil was frequently allowed to cause a "gusher." When this happened, a jet of oil was shot several hundred feet into the air and wasted.

147

RESIDUAL SOILS are those formed in place by the gradual decay of parent material. When cut through they show a gradual transition from fresh rock up to decayed rock to subsoil and topsoil. Residual soils usually form slowly. The deepening soil layer protects the rock beneath from further chemical action. Types of residual soils depend on the rock from which they form, climate, and other factors.

SOIL is the best known, most complex rock and, fortunately, one which has been studied a long time. The physical and chemical weathering of surface rock (parent material) with the addition of organic material, forms soil. Plants and animals (especially microscopic forms) contribute greatly to soil formation, as do climate, vegetation, time, slope, and drainage. The result of centuries of activity is a soil mantle from a few inches to a hundred feet thick, though the average depth of soil is only a foot or so. Since most life depends on soils, they should be preserved and skillfully managed.

TRANSPORTED SOILS are developed on parent material that has been moved by wind, water, or ice. Huge deposits of wind-blown silt serve as the parent material for the loess soils in the western prairies. In the Mississippi valley and along western streams are deposits of water-borne alluvial materials on which some very rich soils have formed. In northern states glacial debris is often the parent material.

TROPICAL RED SOILS are well-developed, well-drained soils resulting from the deep leaching action of much rain and the chemical action of warm air. These may be residual soils of great thickness. The leaching and oxidation make poor soil which may be exhausted after a few years of cultivation. The fall of the Mayan civilizations has been attributed to soil exhaustion.

Soils may be classified in a half dozen ways, according to various properties. They may be classified on the basis of texture (size of particle), as clayey, silty, or sandy. Other classifications have been based on color, parent material, type of crop raised, and many other bases. Most modern classifications begin with three great divisions or zones ranging from immature to mature soils. Three examples of well-developed soils are (1) the tropical red soils (true and modified laterites), (2) the northern forest soils (podsols) and their modifications, and (3) the grassland soils (chernozem and prairie soils).

NORTHERN FOREST SOILS illustrate different conditions from those above. These gray soils form under beds of spruce, pine, and fir needles which are acid in composition and decay slowly. The organic and inorganic materials mix poorly. Many of these soils of cooler regions have not been altered enough to make them good producers without special handling.

RED CLAY
Missouri

GRAY CLAY
Illinois

CLAY was used by primitive men to make pottery not long after they first began to use stones as tools and weapons. After centuries of service, clay is still essential in many industries. About 60 million tons are mined annually and are used in bricks, pottery, chinaware, ceramic pipe, drilling muds, and for many other purposes. Clays are a mixture of silica, alumina, and water. Clay particles are small—less than 0.0001 inch. They stick together but are slippery when moist. Clays may come from granitic rocks, as a weathering product of the feldspars. They also form from weathered shales which came mainly from clay minerals originally. Clay deposits form on lake bottoms and in other quiet water, sometimes with annual layers (varves). One of the clay minerals is kaolin (p. 49).

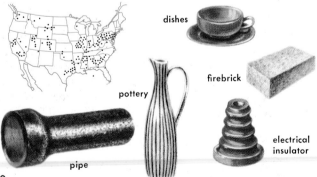

dishes

firebrick

pottery

electrical
insulator

pipe

150

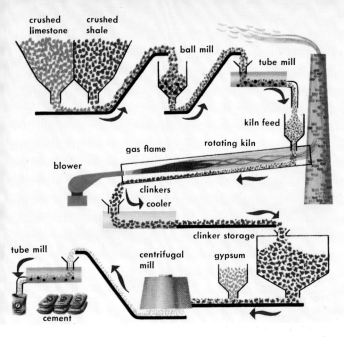

PORTLAND CEMENT is the best-known ingredient of artificial stone. It has largely replaced the older natural cement or hydraulic limestone. When limestone containing silica and clay was burned, the lime that formed would set under water, hence the name hydraulic lime.

Now, limestone and shale are crushed, dried, mixed in the correct proportions, and ground to a fine powder. The powdered mixture is burned in a sloping rotary kiln at about 2700°F to form a glassy clinker. The clinker is crushed, a small amount of gypsum is added, and the mixture is reground to form cement. Over 200 million barrels of Portland cement are produced each year. Cement is mixed with sand, crushed rock, and water to make concrete.

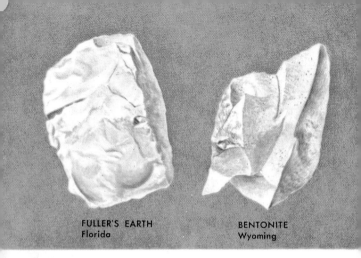

FULLER'S EARTH
Florida

BENTONITE
Wyoming

FULLER'S EARTH is clay or a silty clay material containing over half silica, valued for its decolorizing properties. Fuller's earth absorbs dark organic matter from fats, oils, and greases. First used to "full" or degrease woolen cloth, fuller's earth is used to bleach mineral and vegetable oils. It has the greasy feel of clay and usually breaks up in water. Color varies from white to yellow, brown, and blue.

BENTONITE, first developed, like fuller's earth, for bleaching, has turned out to have even more important uses. It is used in soaps and washing compounds and is added to clays to increase their plasticity. Some bentonite, used as an aid in well drilling, expands and seals off water-bearing sands. Other varieties swell little when wet. Bentonite is also used for paper filler and in adhesives. It is a mixture of at least two aluminum and magnesium silicate minerals, and is usually regarded as weathered volcanic ash.

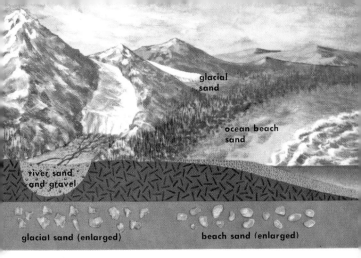

glacial sand

ocean beach sand

river sand and gravel

glacial sand (enlarged)

beach sand (enlarged)

SAND, GRAVEL, AND CRUSHED ROCK have about the lowest value per ton of any rock or mineral, yet all three are indispensable in modern construction. Sand in places is so plentiful that it shapes the landscape, giving beaches and deserts a quiet beauty of their own. Sand is a size term; although sand is usually composed largely of quartz, pure quartz sand (used in the manufacture of glass) is rare. On tropical shores, coral (lime) sands are common. Gypsum sands make up the snowy dunes in the White Sands National Monument. Other sands are rich in magnetite, monazite, garnet, ilmenite, and rutile. Some are mined as ores. But it is common quartz sand that helps make our concrete roads, bridges, and buildings.

Gravel, of glacial or stream origin, contains larger pebbles from a quarter of an inch up, cobbles, and even boulders, in a sand matrix. Washed, screened, and sorted, gravel is used for fill and in concrete. Limestone, basalt, and granite are crushed for road building, railroad ballast, and concrete work.

BUILDING STONES are those cut to size for buildings or monuments. Ornamental stones are those used for finishing or decoration. Stone also has been used for sidewalks, curbing, and paving blocks. For building use, ease of quarrying, transportation, durability, color,

RED GRANITE
Wausau, Wis.

GRANITE is famous for its beauty, strength, and durability —hence its wide use in monuments and buildings. Granite takes a high polish and is resistant to weathering. Its hardness and lack of bedding make quarrying difficult. Granites for building uses are classified by grain size—with preference for fine-grained rock.

COARSE GABBRO
Aberdeen, Scotland

TRAPROCK is the quarryman's term for diabase, basalt, or gabbro. These hard, durable rocks are limited in building use because their iron minerals give a rusty stain as they weather. Traprock is excellent as crushed rock and is widely used. Other igneous rocks—rhyolites and felsites —are locally used as building stones.

"CRAB ORCHARD" STONE
Tennessee

SANDSTONE is relatively easy to quarry because it is bedded. Many sandstones are attractive and durable, and once were very fashionable in Eastern cities. Porous sandstone may not weather well and special treatment may be needed in cold regions. Otherwise, sandstones make attractive building stones. Color and texture are variable.

weathering characteristics, and freedom from iron minerals are important. Of the building stones which are still widely used, the following general types are the best known. Many different kinds of marbles, limestones, and granites are used for varied effects.

LIMESTONE from near Bedford in central Indiana is a well-known building stone, widely used for public buildings. Bedford limestone is white, even-textured and oolitic, sometimes packed with small fossils. It is easily quarried and uniform in color. Many other kinds of limestones are used for building in many parts of the United States.

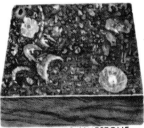
FOSSILIFEROUS LIMESTONE
Swanton, Vt.

MARBLE for building use also includes fine-grained ornamental limestones which are not true marbles. Marble is a classic stone, worked by sculptors as well as builders. Italian white and Belgian black marble are famous Marbles may also be pink, yellow, and brown. They are softer and less resistant to weathering than granites.

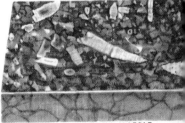
BROWN MARBLE
Killarney, Ireland

SLATE is an unusual building stone used primarily in roofing and flooring, blackboards, and electrical switchboards. It breaks along cleavage planes in large flat sheets. Color varies from black to green and red. Slate is durable and attractive as roofing and in floors and patios. Most U.S. slate comes from Vermont, Maine, and Pennsylvania.

GRAY SLATE   RED SLATE
Bangor, Pa.   New York

155

MORE INFORMATION

There are many sources of information that will help you find out more about rock and minerals. In addition to those here, check on state or provincial publications and those of the U.S. Geological Survey.

Books

The following are a small sample of books available on mineralogy. Several other titles of interest can be found on p. 91.

Sorrell, Charles A. and George F. Sandstrom, **The Golden Field Guide to Rocks and Minerals,** St. Martin's Press, New York, 2000. A companion to this book, it goes into more detail.

Pough, Frederick H., **A Field Guide to Rocks and Minerals,** 5th edition, Houghton Mifflin Co., Boston, 1998.

Hurlbut, Cornelius S., Jr. and W. Edwin Sharp, **Dana's Minerals and How to Study Them,** 4th ed., John Wiley & Sons, New York, 1998.

The Internet

There are many Internet sites devoted to minerals. Some keywords to use in searches include "mineralogy," "mineral specimens," mineralogy of a state or locality of your choice (for example, "mineralogy of Colorado"), or the mineralogy of a particular element (such as the "mineralogy of beryllium").

Magazines

There are two very good magazines available:

Rock and Minerals, Heldref Publications, 1319 18th Street NW, Washington, DC 20078-6117. The bimonthly magazine has articles on all aspects of earth science.

The Mineralogical Record, P.O. Box 35565, Tucson, AZ 85740. Published bimonthly, this magazine is for the advanced mineral collector.

MUSEUMS AND EXHIBITS will show you more kinds of rocks and minerals than you will find on field trips. Use one to supplement the other. Some of the museums with larger exhibits are listed below.

Arizona
Holbrook: Petrified Forest National Monument Museum; Tucson: University of Arizona

California
Berkeley: University of California; Los Angeles: Los Angeles County Museum; San Diego: Natural History Museum; San Francisco: California State Division of Mines Museum; Santa Barbara: Santa Barbara Museum of Natural History

Colorado
Boulder: University of Colorado Museum; Denver: Denver Museum of Natural History

Connecticut
New Haven: Peabody Museum of Natural History

Georgia
Atlanta: Georgia State Museum

Illinois
Chicago: Field Museum of Natural History; Urbana: University of Illinois Museum of Natural History

Indiana
Indianapolis: Indiana State Museum

Kansas
Lawrence: University of Kansas Museum

Louisiana
Baton Rouge: State University Department of Geology Museum

Massachusetts
Cambridge: Harvard University Mineralogical Museum; Springfield: Museum of Natural History

Michigan
Ann Arbor: University of Michigan Mineralogy Museum; Bloomfield Hills: Cranbrook Institute of Science; Houghton: Michigan College of Mining and Technology

Missouri
Columbia: University of Missouri Museum; Jefferson City: Missouri Resources Museum

Montana
Butte: Montana School of Mines

New Jersey
Paterson: Paterson Museum; Trenton: New Jersey State Museum

New Mexico
Socorro: New Mexico Institute of Mining and Technology

New York
Albany: New York State Museum; Buffalo: Buffalo Museum of Science; New York: American Museum of Natural History

Ohio
Cleveland: Cleveland Museum of Natural History

Pennsylvania
Philadelphia: Philadelphia Academy of Natural Sciences; Pittsburgh: Carnegie Museum

South Dakota
Rapid City: South Dakota School of Mines and Technology

Washington, D.C.
Smithsonian Museum of Natural History

Ontario
Ottawa: National Museum of Canada; Toronto: Royal Ontario Museum

Quebec
Montreal: Redpath Museum

INDEX

Asterisks (*) indicate illustrations.

159

MEASURING SCALE (IN 10THS OF AN INCH)